DOMAIN THEORY, LOGIC AND COMPUTATION

Semantic Structures in Computation

Volume 3

Title of the Series:
Semantic Structures in Computation

Series Editor in Chief
Guo-Qiang Zhang
Case Western Reserve University
Department of Electrical Engineering and Computer Science
Olin 610, 10900 Euclid Avenue, Cleveland, OH 44106, USA
E-mail: gqz@eecs.cwru.edu

Scope of the Series
With the idea of partial information and approximation as the starting point, this bookseries focuses on the interplay among computer science, logic, and mathematics through algebraic, order-theoretic, topological, and categorical means, with the goal of promoting cross-fertilization of ideas and advancing interdisciplinary research.

This bookseries provides a distinctive publication forum for collected works and monographs on topics such as domain theory, programming semantics, types, concurrency, lambda calculi, topology and logic in computer science, and especially applications in non-traditional and emerging areas in which the development of formal semantics deepens our understanding of a computational phenomenon.

The titles published in this series are listed at the end of this volume.

DOMAIN THEORY, LOGIC AND COMPUTATION

Proceedings of the 2nd International Symposium on Domain Theory, Sichuan, China, October 2001

Edited by

GQ Zhang
Case Western Reserve University, Cleveland, U.S.A.

J. Lawson
Louisiana State University, Baton Rouge, U.S.A.

Y.-M. Liu
Sichuan University, Sichuan, China

and

M.-K. Luo
Sichuan University, Sichuan, China

SPRINGER-SCIENCE+BUSINESS MEDIA, B.V.

A C.I.P. Catalogue record for this book is available from the Library of Congress.

ISBN 978-90-481-6523-0 ISBN 978-94-017-1291-0 (eBook)
DOI 10.1007/978-94-017-1291-0

Printed on acid-free paper

Contents

Preface

This volume is devoted to the Proceedings of the 2nd International Symposium on Domain Theory (ISDT 2001) held from October 22 to 26, 2001, at Sichuan University in Chengdu, China.

Domains emerged in the late nineteen sixties and early seventies in the seminal work of Scott on semantics of programming languages and lambda calculi as well as in the work of Ershov in his fundamental investigations on computable functionals of higher types. The subject of domain theory as we know today is founded by Scott and Plotkin. Domains are mathematical structures modelling the order of information contained in computation by approximating ideal objects by finite ones. Thus they combine order-theoretical and topological structures and provide a natural framework for modelling and reasoning about computation. The theory of domains has seen a rapid development. Not only has it proved to be a useful tool for the semantics of programming languages, in particular functional languages, it also has led to interesting developments and applications in mathematics and other areas of computer science.

In China, the research in theory of domains has been developed since the nineteen eighties and has gained a surprising standard. Because of the lack of communication, this research did not become well-known in Europe and in North America before the first International Symposium on Domain Theory (ISDT 1999) successfully held in China in 1999. During the first symposium the participants from China, Great Britain, Germany, Italy, Norway, and U.S.A. reached a consensus to continue similar international symposia on domain theory every two years as a series. ISDT 2001 was the second international conference with the aim to provide an opportunity for researchers in China and in the world to join together for exchanging ideas and results of ongoing research, and to promote research in this interdisciplinary field. We are pleased to see these goals being realized as a result of the contact established by this conference series, as evidenced by the fact that Professor J. Lawson of U.S.A. and Professor P.-L. Curien of France have both visited Sichuan

University and other institutions in China, Professor K. Keimel of Germany has visited Shanghai Normal University, Professor G.-Q. Zhang has visited Chinese Academy of Sciences, Shaanxi Normal University, and Shanghai Normal University. In the reverse direction, Professor G.-J. Wang has participated in the Dagstuhl Seminar on Mathematical Structures for Computable Topology and Geometry and Professor Y.-X. Chen in the Domain Workshop at Birmingham, England. Professor M.-K. Luo conducted a research visit in the U.S.A. at the home universities of Professors J. Lawson and G.-Q. Zhang, and Professor L.-S. Xu at the home university of Professor J. Lawson.

The program committee consisted of

Ying-Ming Liu, Chair, Sichuan University, China
Mao-Kang Luo, Co-chair, Sichuan University, China
Jimmie Lawson, Co-chair, Louisiana State University, U.S.A.
Guo-Qiang Zhang, Co-chair, Case Western Reserve University, U.S.A.

who also serve as the editors of this volume. The conference was intended to be centered around the following themes: topological and logical aspects of domains, categories of domains and powerdomains, partial orders and metric spaces, applications in databases, mathematics, and AI, types, process algebra and concurrency, non-classical and partial logics, programming language semantics.

The meeting was attended by thirty three participants from China and fourteen from other countries such as France, Germany, Great Britain, Ireland, and U.S.A. The scientific program of the symposium was divided into seven invited lectures and twenty four session lectures. In the lectures, a great variety of topics were covered.

In the closing ceremony it was commonly agreed that the conference had fully succeeded in its goals and the next conference would be held in Xi'an, China. The conference should be centered around domain theory and its applications; it should stress the links with computer science, logic, and semantics; and it should ideally have a particular focus on a specific topic resulting from new developments and applications.

Included in this proceedings are selected papers of original research presented at the symposium, submitted to the editors after the conference and accepted after a rigorous refereeing process. The nine published papers are representative for the themes of the workshop. A few of them, such as the contributions by P.-L. Curien and J. Longley, have a survey character. The articles have been grouped according to these topics: domains and computation, topology and convergence, domains, lattices, and continuity, and representations of domains as event and logical structures.

Acknowledgements. The Symposium has been sponsored by Sichuan University, the National Natural Science Foundation of China, and the Mathematical Center of the Educational Ministry of China. It has been supported by the Fuzzy Mathematics and Systems Association of the Systems Engineering Society of China. This Proceedings volume owes its existence to the support of these organizations for the symposium.

Thanks are due to the staff personnel of the Mathematical College of Sichuan University for their local arrangements which made the symposium an enjoyable experience.

The editors of this volume are indebted to the referees for their careful reading, their advice, and their valuable suggestions. These have contributed to the quality of this publication.

Finally, the continued support and suggestions from Charles Erkelens, Senior Publishing Editor at Kluwer Academic Publishers, are greatly appreciated.

YING-MING LIU, MAO-KANG LUO, JIMMIE LAWSON, GUO-QIANG ZHANG

Contributing Authors

Felipe Bracho is a Professor at the Universidad Nacional Autónoma de México, México
bracho@servidor.unam.mx

Pierre-Louis Curien is a Professor at CNRS & Université Paris 7, Paris, France
curien@pps.jussieu.fr

Manfred Droste is a Professor at Institut für Algebra, Technische Universität Dresden, Dresden, Germany
droste@math.tu-dresden.de

Ralf Greb is a Researcher at Theoretische Informatik, Fachbereich Mathematik, Universität Siegen, Siegen, Germany
ralfgreb@informatik.uni-siegen.de

Roland Heinze is a Professor at Institut für Informatik III, Rheinische Friedrich-Wilhelms-Universität Bonn, Bonn, Germany
heinze@nikocity.de

Pascal Hitzler is a Post-doctoral Researcher at Artificial Intelligence Institute, Dresden University of Technology, Dresden, Germany
phitzler@inf.tu-dresden.de

Hui Kou is a Lecturer at Department of Mathematics, Sichuan University, Chengdu, China
kouhui@vip.sina.com

Yong-Ming Li is a Professor at Department of Mathematics, Shaanxi Normal University, Xi'an, China
liyongm@snnu.edu.cn

Ying-Ming Liu is a Professor at Department of Mathematics, Sichuan University, Chengdu, China
ymliu@scu.edu.cn

John R. Longley is a Lecturer at the School of Informatics, University of Edinburgh, Edinburgh, U.K.
jrl@informatics.ed.ac.uk

Mao-Kang Luo is a Professor at Department of Mathematics, Sichuan University, Chengdu, China
makaluo@scu.edu.cn

Ingmar Meinecke is a Research Student at the Institut für Algebra, Technische Universität Dresden, Dresden, Germany
meinecke@math.tu-dresden.de

Anthony Karel Seda is a Senior Lecturer at the Department of Mathematics, University College Cork, Cork, Ireland
a.seda@ucc.ie

Dieter Spreen is a Professor at the Theoretische Informatik, Fachbereich Mathematik, Universität Siegen, Siegen, Germany
spreen@informatik.uni-siegen.de

Xiao-Quan Xu is a Professor at the Department of Mathematics, Sichuan University, and Department of Mathematics, Jiangxi Normal University, China
xiqxu@public.nc.jx.cn

Meng Zhou is a Professor at the Department of Applied Mathematics, Beijing University of Aeronautics and Astronautics, Beijing, China.

Chapter 1

PLAYFUL, STREAMLIKE COMPUTATION

Pierre-Louis Curien
PPS
CNRS & Université Paris 7
Case 7014, 2 pl. Jussieu, 75251 Paris Cedex 05, France
curien@pps.jussieu.fr

Abstract We offer a short tour into the interactive interpretation of sequential programs. We emphasize *streamlike* computation – that is, computation of successive bits of information upon request. The core of the approach surveyed here dates back to the work of Berry and the author on *sequential algorithms on concrete data structures* in the late seventies, culminating in the design of the programming language CDS, in which the semantics of programs of any type can be explored interactively. Around one decade later, two major insights of Cartwright and Felleisen on one hand, and of Lamarche on the other hand gave new, decisive impulses to the study of sequentiality. Cartwright and Felleisen observed that sequential algorithms give a direct semantics to control operators like `call-cc` and proposed to include explicit errors both in the syntax and in the semantics of the language PCF. Lamarche (unpublished) connected sequential algorithms to linear logic and games. The successful program of games semantics has spanned over the nineties until now, starting with syntax-independent characterizations of the term model of PCF by Abramsky, Jagadeesan, and Malacaria on one hand, and by Hyland and Ong on the other hand.

Only a basic acquaintance with λ-calculus, domains and linear logic is assumed in sections 1 through 3.

Keywords: Coroutines, sequentiality, games, abstract machines, Böhm trees

1. Prologue: playing with Böhm trees

We first make some preparations. For self-containedness, we briefly recall the relevant notions. The syntax of the untyped λ-calculus (λ-calculus for short) is given by the following three constructions: a *vari-*

GQ Zhang et al. (eds.), Domain Theory, Logic and Computation, 1-24.

able x is a λ-term, if M and N are λ-terms, then the *application* MN is a λ-term, and if M is a term then the *abstraction* $\lambda x.M$ is a term. Usual abbreviations are $\lambda x_1 x_2.M$ for $\lambda x_1.(\lambda x_2.M)$, and MN_1N_2 for $(MN_1)N_2$, and similarly for n-ary abstraction and application. A more macroscopic view is quite useful: it is easy to check that any λ-term has exactly one of the following two forms:

$$\begin{array}{ll} (n \geq 1, p \geq 1) & \lambda x_1 \cdots x_n.xM_1 \cdots M_p \\ (n \geq 0, p \geq 1) & \lambda x_1 \cdots x_n.(\lambda x.M)M_1 \cdots M_p \ . \end{array}$$

The first form is called a *head normal form* (hnf), while the second exhibits the *head redex* $(\lambda x.M)M_1$. The following easy property justifies the name of head normal form: any reduction sequence starting from a hnf $\lambda x_1 \cdots x_n.xM_1 \cdots M_p$ consists of an interleaving of independent reductions of $M_1, \ldots, M_p$. More precisely, we have:

$$\begin{array}{l} (\lambda x_1 \cdots x_n.xM_1 \cdots M_p \rightarrow^* P) \Rightarrow \\ \exists N_1, \ldots N_p \left\{ \begin{array}{l} P = \lambda x_1 \cdots x_n.xN_1 \cdots N_p \text{ and} \\ \forall i \leq p \ M_i \rightarrow^* N_i \ . \end{array} \right. \end{array}$$

Here, reduction means the replacement in any term of a sub-expression of the form $(\lambda x.M)N$, called a β-redex, by $M[x \leftarrow N]$. A normal form is a term that contains no β-redex, or equivalently that contains no head redex. Hence the syntax of normal forms is given by the following two constructions: a variable x is a normal form, and if $M_1, \ldots, M_p$ are normal forms, then $\lambda x_1 \cdots x_n.xM_1 \cdots M_p$ is a normal form.

Now, we are ready to play. Consider the following two normal forms:

$$M = zM_1M_2(\lambda z_1 z_2.z_1 M_3 M_4) \qquad N = \lambda x_1 x_2 x_3.x_3(\lambda y_1 y_2.y_1 N_1)N_2 \ .$$

The term $M[z \leftarrow N] = NM_1M_2(\lambda z_1 z_2.z_1 M_3 M_4)$ is not a normal form anymore, and can be β-reduced as follows:

$$\begin{array}{rcl} NM_1M_2(\lambda z_1 z_2.z_1 M_3 M_4) & \rightarrow & (\lambda z_1 z_2.z_1 M_3 M_4)(\lambda y_1 y_2.y_1 N_1')N_2' \\ & \rightarrow & (\lambda y_1 y_2.y_1 N_1')M_3'M_4' \\ & \rightarrow & M_3'N_1'' \ , \end{array}$$

where N_1', etc... are suitable substitution instances of N_1 etc... But there is a more geometric way of describing the *interaction* of M and N. First, we represent M and N explicitly as trees (read from left to right), as follows:

$$z \left\{ \begin{array}{l} M_1 \\ M_2 \\ \lambda z_1 z_2. \ z_1 \left\{ \begin{array}{l} M_3 \\ M_4 \end{array} \right. \end{array} \right. \qquad \lambda x_1 x_2 x_3. \ x_3 \left\{ \begin{array}{l} \lambda y_1 y_2. \ y_1 \left\{ \ N_1 \right. \\ N_2 \end{array} \right.$$

Then we represent computation as the progression of two tokens in the two trees. Initially, the tokens are at the root (we use underlining to indicate the location of the tokens):

$$\underline{z}\left\{\begin{array}{l} M_1 \\ M_2 \\ \lambda z_1 z_2.\ z_1 \left\{\begin{array}{l} M_3 \\ M_4 \end{array}\right. \end{array}\right. \qquad \underline{\lambda x_1 x_2 x_3.}\ x_3 \left\{\begin{array}{l} \lambda y_1 y_2.\ y_1 \left\{\ N_1\right. \\ N_2 \end{array}\right.$$

We then notice that the token in M has a choice of where to move right, while the one on the right has no choice. So let us take the latter as pilot:

$$\underline{z}\left\{\begin{array}{l} M_1 \\ M_2 \\ \lambda z_1 z_2.\ z_1 \left\{\begin{array}{l} M_3 \\ M_4 \end{array}\right. \end{array}\right. \qquad \lambda x_1 x_2 x_3.\ \underline{x_3} \left\{\begin{array}{l} \lambda y_1 y_2.\ y_1 \left\{\ N_1\right. \\ N_2 \end{array}\right.$$

Here, $\underline{x_3}$ reads as "take the third branch" (in M):

$$z\left\{\begin{array}{l} M_1 \\ M_2 \\ \underline{\lambda z_1 z_2.}\ z_1 \left\{\begin{array}{l} M_3 \\ M_4 \end{array}\right. \end{array}\right. \qquad \lambda x_1 x_2 x_3.\ \underline{x_3} \left\{\begin{array}{l} \lambda y_1 y_2.\ y_1 \left\{\ N_1\right. \\ N_2 \end{array}\right.$$

Notice that we are now in a situation where the token in M knows how to move next, while the one in N has a choice. We take M as pilot:

$$z\left\{\begin{array}{l} M_1 \\ M_2 \\ \lambda z_1 z_2.\ \underline{z_1} \left\{\begin{array}{l} M_3 \\ M_4 \end{array}\right. \end{array}\right. \qquad \lambda x_1 x_2 x_3.\ \underline{x_3} \left\{\begin{array}{l} \lambda y_1 y_2.\ y_1 \left\{\ N_1\right. \\ N_2 \end{array}\right.$$

We read z_1 as "take the first branch" (in N):

$$z\left\{\begin{array}{l} M_1 \\ M_2 \\ \lambda z_1 z_2.\ \underline{z_1} \left\{\begin{array}{l} M_3 \\ M_4 \end{array}\right. \end{array}\right. \qquad \lambda x_1 x_2 x_3.\ x_3 \left\{\begin{array}{l} \underline{\lambda y_1 y_2.}\ y_1 \left\{\ N_1\right. \\ N_2 \end{array}\right.$$

The following steps are:

$$z \left\{ \begin{array}{l} M_1 \\ M_2 \\ \lambda z_1 z_2.\ \underline{z_1} \left\{ \begin{array}{l} M_3 \\ M_4 \end{array} \right. \end{array} \right. \qquad \lambda x_1 x_2 x_3.\ x_3 \left\{ \begin{array}{l} \lambda y_1 y_2.\ \underline{y_1} \left\{ \ N_1 \right. \\ N_2 \end{array} \right.$$

$$z \left\{ \begin{array}{l} M_1 \\ M_2 \\ \lambda z_1 z_2.\ z_1 \left\{ \begin{array}{l} \underline{M_3} \\ M_4 \end{array} \right. \end{array} \right. \qquad \lambda x_1 x_2 x_3.\ x_3 \left\{ \begin{array}{l} \lambda y_1 y_2.\ \underline{y_1} \left\{ \ N_1 \right. \\ N_2 \end{array} \right.$$

We leave it to the reader to check that these steps follow closely the sequence of β-reductions given above. The graphical presentation enhances alternation. The tokens' moves alternate between M and N. There are two sorts of moves: variables (like z), and (bunches of) abstractions (like $\lambda z_1 z_2.$). We call these moves Player's moves and Opponent's moves, respectively. We can view an Opponent's move as the question "what is the head variable of the term rooted here?", and a Player's move as the answer to this question. So we see the computation as a progression of alternating moves describing a path in M (and in N):

$$\underline{z} \left\{ \begin{array}{l} M_1 \\ M_2 \\ \underline{\lambda z_1 z_2.}\ \underline{z_1} \left\{ \begin{array}{l} \underline{M_3} \\ M_4 \end{array} \right. \end{array} \right. \qquad \underline{\lambda x_1 x_2 x_3.}\ \underline{x_3} \left\{ \begin{array}{l} \underline{\lambda y_1 y_2.}\ \underline{y_1} \left\{ \ N_1 \right. \\ N_2 \end{array} \right.$$

Our example is actually too simple. The general mechanism needs an explicit manipulation of pointers, when (unlike in the example) a variable is not bound by the immediate bunch of λ's above. We refer the interested reader to [15, 16], where this machinery is described for a larger class of trees with pointers – called abstract Böhm trees –, of which Böhm trees are an example. Our main point here was to highlight interaction: M and N are pilot in turn and tell the other which branch to choose.

Suppose now that $M_3 = \lambda u.tM_5M_6$, where t is a free variable of $M[z \leftarrow N]$. Then, looking back at the sequence of β-reductions, we reach a head normal form:

$$M_3' N_1'' = (\lambda u.tM_5' M_6') N_1'' \quad \rightarrow \quad tM_5'' M_6''\ .$$

And, in geometrical form:

$$z\left\{\begin{array}{l} M_1 \\ M_2 \\ \lambda z_1 z_2.\, z_1 \left\{\begin{array}{l} \lambda u.\, \underline{t}\left\{\begin{array}{l} M_5 \\ M_6 \end{array}\right. \\ M_4 \end{array}\right. \end{array}\right. \qquad \lambda x_1 x_2 x_3.\, x_3 \left\{\begin{array}{l} \lambda y_1 y_2.\, \underline{y_1}\left\{\ N_1 \right. \\ N_2 \end{array}\right.$$

Note here that N cannot help to choose the next move in M. The machinery stops here. After all, most functional programming languages stop evaluation on (weak) head normal forms. But what about getting the full normal form, i.e., computing M_5'' and M_6''? The interactive answer to this question is: by exploration of branches, on demand, or in a streamlike manner. The machine displays t as the head variable of the normal form of $M[z \leftarrow N]$. Now, you, the Opponent, can choose which of the branches below t to explore, and then the machine will restart until it reaches a head normal form. For example, if you choose the first branch, then you will eventually reach the head variable of M_5''. This is called streamlike, because that sort of mechanism has been first analysed for infinite lists built progressively. A λ-term too has a "potentially infinite normal form": it's Böhm tree.

This prologue served the purpose of introducing some keywords, such as interactivity, playful interpretation, streamlike computation. We now start from the beginning.

2. Introduction

Scott's and Plotkin's denotational semantics takes its roots in recursion theory. It is worth recalling here the statement of Rice's theorem. This theorem asserts a property of recursively enumerable (r.e.) sets of partial recursive (p.r.) functions, defined through a fixed enumeration (ϕ_n) of the p.r. functions (i.e. ϕ is a surjection from ω – the set of natural numbers – to $\omega \rightharpoonup \omega$, using $\rightharpoonup$ for sets of partial functions). Let $PR \subseteq \omega \rightharpoonup \omega$ denote the set of p.r. functions. A subset $A \subseteq PR$ is called r.e. if $\{n \mid \phi_n \in A\}$ is r.e. in the usual sense. The theorem asserts that if A is r.e. and if $f \in A$, then there exists a finite approximation g of f such that $g \in A$. That g is an approximation of f means that f is an extension of g, i.e., the domain on which the partial function f is defined, or domain of definition of f, contains that of g and f and g coincide where they are both defined. A simpler way of saying this is that the graph of g is contained in the graph of f. Moreover, the domain of definition of g is finite. Rice's theorem is about an intrinsic continuity property in the realm of p.r. functions. It highlights the (complete) partial order structure of $\omega \rightharpoonup \omega$, and in particular the presence of a bottom element $\bot$ in this partial order: the everywhere undefined function.

Certainly, one of the key departure points taken by Scott was to take $\bot$ seriously. Once this element is part of the picture, one takes a new look at some basic functions. Take the booleans, for example. In Scott's semantics, this is not the set $\{tt, ff\}$, but the set $\{\bot, tt, ff\}$ ordered as follows: $x \leq y$ if and only if $x = y$ or $x = \bot$ (this is called *flat* ordering). Take now the good old disjunction function $or : Bool \times Bool \to Bool$. It gives rise to four different functions over the flat domain version of *Bool* (the specifications below can be completed to full definitions by monotonicity):

$$\begin{aligned} por(\bot, tt) &= tt \\ por(tt, \bot) &= tt \\ por(\bot, ff) &= \bot \\ por(ff, \bot) &= \bot \\ por(ff, ff) &= ff \end{aligned}$$

$$\begin{aligned} lor(\bot, y) &= \bot \\ lor(tt, \bot) &= tt \\ lor(ff, y) &= y \end{aligned} \qquad \begin{aligned} ror(\bot, tt) &= tt \\ ror(x, \bot) &= \bot \\ ror(x, ff) &= x \end{aligned}$$

$$\begin{aligned} sor(\bot, tt) &= \bot \\ sor(tt, \bot) &= \bot \\ sor(\bot, ff) &= \bot \\ sor(ff, \bot) &= \bot \\ sor(ff, tt) &= tt \\ sor(tt, ff) &= tt \\ sor(tt, tt) &= tt \\ sor(ff, ff) &= ff \end{aligned}$$

It should be clear that *lor* and *ror* are computed by programs of the following shape, respectively:

$$\lambda xy.\mathtt{if}\, x = tt \begin{cases} \mathtt{then}\; tt \\ \mathtt{else\;if}\; y = \cdots \end{cases} \qquad \lambda xy.\mathtt{if}\, y = tt \begin{cases} \mathtt{then}\; tt \\ \mathtt{else\;if}\; x = \cdots \end{cases}$$

On the other hand, it should be intuitively clear that no sequential program of the same sort can compute *por*, because a sequential program will

- either start by examining one of the arguments, say x, in which case it can't output anything before a value for x is given, thus missing the specification $por(\bot, tt) = tt$,

- or output some value rightaway, say tt ($\lambda xy.tt$), thus mising the specification $por(\bot, \bot) = \bot$.

For a formal proof that *por* is not sequentially definable, we refer to [33] (syntactic proof), to [22][section 6.1] (model-theoretic proof), and to [5][section 4.5] (via logical relations). As for *sor*, the story is yet different, there are *two* natural sequential programs for it:

$$\lambda xy.\mathtt{if}\, x = tt \left\{ \begin{array}{l} \mathtt{then\,if}\; y = \cdots \\ \mathtt{else\,if}\; y = \cdots \end{array} \right. \qquad \lambda xy.\mathtt{if}\, y = tt \left\{ \begin{array}{l} \mathtt{then\,if}\; x = \cdots \\ \mathtt{else\,if}\; x = \cdots \end{array} \right.$$

The starting point of the model of sequential algorithms (next section) was to interpret these two programs as different objects *lsor* and *rsor*. Notice finally that there are many more sequential programs computing *lor*, *ror*, or *sor*. Another program for *lor* might e.g. look like

$$\lambda xy.\ \mathtt{if}\ x = tt \left\{ \begin{array}{l} \mathtt{then}\ tt \\ \mathtt{else\ if}\ x = tt \left\{ \begin{array}{l} \mathtt{then}\ tt \\ \mathtt{else\ if}\ y = \cdots \end{array} \right. \end{array} \right.$$

Such a "stuttering" program is perfectly correct syntactically. Whether this program is interpreted in the model by an object different from the above program for *lor* is the departure point between the model of sequential algorithm on one hand and the more recent games semantics on the other hand. We shall come back to this point in the next section.

Before we close the section, let us give some rationale for the names used in this section. As the reader might have guessed, the prefixes p, l, r, s, ls, rs stand for "parallel", "left", "right", "left strict", and "right strict", respectively.

3. Symmetric algorithms, sequential algorithms

We introduce enough formal notions to give a self-contained introduction to sequential algorithms, presented in the light of a games interpretation [12] (following work of Lamarche [27]). The proofs are omitted, but can be found in [5][section 14.3], except for what regards the coincidence between the two definitions of composition, for which the proof from [14][section 3.6] can easily be adapted.

Definition 3.1. A *sequential data structure* $\mathbf{S} = (C, V, P)$ is given by two sets C and V of *cells* and *values*, which are assumed disjoint, and by a collection P of non-empty words p of the form:

$$c_1v_1 \cdots c_nv_n \text{ or } c_1v_1 \cdots c_{n-1}v_{n-1}c_n,$$

where $c_i \in C$ and $v_i \in V$ for all i. Thus any $p \in P$ is alternating and starts with a cell. Moreover, it is assumed that P is closed under non-empty prefixes. We call the elements of P *positions* of $\mathbf{S}$. We call *move* any element of $M = C \cup V$. We use m to denote a move. A position ending with a value is called a *response*, and a position ending with a cell is called a *query*. We use p (or s, or t), q, and r, to range over positions, queries, and responses, respectively. We denote by Q and R the sets of queries and responses, respectively.

Let us pause here for some comments and perspective. An important step in the semantic account of sequential computing was taken by Berry, who developed the stable model in which the function *por* is excluded. Winskel described this model more concretely in terms of *event structures*, and Girard proposed a simpler form called *coherence spaces*, that led him to the discovery of linear logic [19] (see also [5][chapters 12 and 13]). In event structures or coherence spaces, data are constructed out of elementary pieces, called *events*, or tokens. For example, the pair of booleans $(tt, f\!f)$ is obtained as the set of two elementary pieces: $(tt, \bot)$ and $(\bot, f\!f)$. More precisely and technically, the structure $Bool \times Bool$ as a coherence space has four events: $tt.1$, $f\!f.1$, $tt.2$, and $f\!f.2$. Then $(tt, f\!f)$ is the set $\{tt.1, f\!f.2\}$.

In a sequential data structure (or in a concrete data structure, not defined here) events are further cut in two "halves": a cell and a value, or an opponent's move and a player's move. The structure $Bool \times Bool$ as an sds has two cells ?.1 and ?.2 and has four values $tt.1$, $f\!f.1$, $tt.2$, and $f\!f.2$. An event, say $tt.1$, is now decomposed as a position (?.1) ($tt.1$). The best way to understand this is to think of a streamlike computation. Your pair of booleans is the output of some program, which will only work on demand. The cell ?.1 reads as "I – another program, or an observer – want to know the left coordinate of the result of the program", and $tt.1$ is the answer to this query.

An important remark, which will be further exploited in section 5, is that this decomposition of events gives additional space: there is no counterpart in the world of coherence spaces or in any other usual category of domains of a structure with one cell and no value.

Definition 3.2. A *strategy* of $\mathbf{S}$ is a subset x of R that is closed under response prefixes and binary non-empty greatest lower bounds (glb's):

$$r_1, r_2 \in x, r_1 \wedge r_2 \neq \epsilon \quad \Rightarrow \quad r_1 \wedge r_2 \in x \ ,$$

where ϵ denotes the empty word. A *counter-strategy* is a non-empty subset of Q that is closed under query prefixes and under binary glb's. We use $x, y, \ldots$ and $\alpha, \beta, \ldots$ to range over strategies and counter-strategies, respectively.

If x is a strategy and if $r \in x, q = rc$ for some c and if there is no v such that $qv \in x$, we write $q \in A(x)$ (and say that q is accessible from x). Likewise we define $r \in A(\alpha)$ for a response r and a counter-strategy α.

Both sets of strategies and of counter-strategies are ordered by inclusion. They are denoted by $D(\mathbf{S})$ and $D^{\perp}(\mathbf{S})$, respectively. We write $\mathcal{K}(D(\mathbf{S}))$ and $\mathcal{K}(D^{\perp}(\mathbf{S}))$ for the sets of *finite* strategies and counter-strategies, respectively. Notice that $D(\mathbf{S})$ has always a minimum element (the empty strategy, written $\emptyset$ or $\perp$), while $D^{\perp}(\mathbf{S})$ has no minimum element in general.

A more geometric reading of the definitions of sds, strategy and counter-strategy is the following. An sds is a labelled forest, where the ancestor relation alternates cells and values, and where the roots are labelled by cells. A strategy is a sub-forest which is allowed to branch only at values. A counter-strategy α is a non-empty subtree which is allowed to branch only at cells.

Let us see what collections of positions form and do not form a strategy in *Bool* $\times$ *Bool*. The set $\{(?.1)\ (tt.1)\ ,\ (?.2)\ (ff.2\})$ (representing (tt, ff)) is a strategy, while $\{(?.1)\ (tt.1)\ ,\ (?.1)\ (ff.1)\}$ is not a strategy. A way to understand this is to say that the cell ?.1 can hold only one value, which is the answer to the question. A strategy consists in having ready determinate answers for the movements of the opponent. If strategies are data, what are counter-strategies? They can be considered as exploration trees, see below.

The pairs cell–value, query–response, and strategy–counter-strategy give to sds's a flavour of symmetry. These pairs are related to other important dualities in programming: input–output, constructor–destructor (see [17]). It is thus tempting to consider the counter-strategies of an sds $\mathbf{S}$ as the strategies of a dual structure $\mathbf{S}^{\perp}$ whose cells are the values of $\mathbf{S}$ and whose values are the cells of $\mathbf{S}$. However, the structure obtained in this way is not an sds anymore, since positions now start with a value. This situation, first analysed by Lamarche [28], is now well-understood since the thesis work of Laurent [29]. We come back to this below.

The following definition resembles quite closely to the dynamics described in section 1.

Definition 3.3 (play). Let $\mathbf{S}$ be an sds, x be a strategy and α be a counter-strategy of $\mathbf{S}$, one of which is finite. We define $x \,|\, \alpha$, called a *play*, as the set of positions p which are such that all the response prefixes of p are in x and all the query prefixes of p are in α.

Proposition 3.1. *Given x and α as in definition 3.3, the play $x \,\big|\, \alpha$ is non-empty and totally ordered, and can be confused with its maximum element, which is uniquely characterized as follows:*

$$\begin{array}{llll} x \,\big|\, \alpha \textit{ is the unique element of } x \cap A(\alpha) & \textit{if } x \,\big|\, \alpha \textit{ is a response} \\ x \,\big|\, \alpha \textit{ is the unique element of } \alpha \cap A(x) & \textit{if } x \,\big|\, \alpha \textit{ is a query}\,. \end{array}$$

Definition 3.4 (winning). Let x and α be as in definition 3.3. If $x \,\big|\, \alpha$ is a response, we say that x wins against α, and we denote this predicate by $x \triangleleft \alpha$. If $x \,\big|\, \alpha$ is a query, we say that α wins against x, and we write $x \triangleright \alpha$, thus $\triangleright$ is the negation of $\triangleleft$. To stress who is the winner, we write:

$$x \,\big|\, \alpha = \begin{cases} x \triangleleft\!\big|\, \alpha & \text{when } x \text{ wins} \\ x \,\big|\!\triangleright \alpha & \text{when } \alpha \text{ wins}\,. \end{cases}$$

The position $x \,\big|\, \alpha$ formalizes the interplay between the player with strategy x and the opponent with strategy α. If $x \,\big|\, \alpha$ is a response, then the player wins since he made the last move, and if $x \,\big|\, \alpha$ is a query, then the opponent wins. Here is a game theoretical reading of $x \,\big|\, \alpha$. At the beginning the opponent makes a move c: his strategy determines that move uniquely. Then either the player is unable to move (x contains no position of the form cv), or his strategy determines a unique move. The play goes on until one of x or α does not have the provision to answer its opponent's move (cf. section 1).

We next define the morphisms between sds's. There are two definitions, a concrete one and a more abstract one. The concrete one is needed since we want the morphisms to form in turn an sds in order to get a cartesian closed category (actually a monoidal closed one, to start with). Accordingly, there will be two definitions of the composition of morphisms. Their equivalence is just what *full abstraction* – that is, the coincidence of operational and denotational semantics – boils down to, once we have tailored the model to the syntax (programs as morphisms) and tailored the syntax to the semantics (like in the language CDS [7]). We start with the concrete way.

Definition 3.5. Given sets $A, B \subseteq A$, for any word $w \in A^*$, we define $w\lceil_B$ as follows:

$$\epsilon\lceil_B = \epsilon \qquad wm\lceil_B = \begin{cases} w\lceil_B & \text{if } m \in A \backslash B \\ (w\lceil_B)m & \text{if } m \in B\,. \end{cases}$$

Definition 3.6. Given two sds's $\mathbf{S} = (C, V, P)$ and $\mathbf{S}' = (C', V', P')$, we define $\mathbf{S} \multimap \mathbf{S}' = (C'', V'', P'')$ as follows. The sets C'' and V'' are disjoint unions:

$$\begin{array}{lcl} C'' & = & \{\textit{request } c' \mid c' \in C'\} \cup \{\textit{is } v \mid v \in V\} \\ V'' & = & \{\textit{output } v' \mid v' \in V'\} \cup \{\textit{valof } c \mid c \in C\}\,. \end{array}$$

P'' consists of the alternating positions s starting with a *request* c', and which are such that:

$$s\lceil_{\mathbf{S}'} \in P', (s\lceil_{\mathbf{S}} = \epsilon \text{ or } s\lceil_{\mathbf{S}} \in P), \text{ and}$$
$$s \text{ has no prefix of the form } s(\mathit{valof}\ c)(\mathit{request}\ c').$$

We often omit the tags *request*, *valof*, *is*, *output*, as we have just done in the notation $s\lceil_{\mathbf{S}} = s\lceil_{C \cup V}$ (and similarly for $s\lceil_{\mathbf{S}'}$).

We call *affine sequential algorithms* (or affine algorithms) from $\mathbf{S}$ to $\mathbf{S}'$ the strategies of $\mathbf{S} \multimap \mathbf{S}'$.

The constraint 'no scc'' can be formulated more informally as follows. Thinking of *valof* c as a call to a subroutine, the principal routine cannot proceed further until it receives a result v from the subroutine.

The identity affine algorithm $id \in D(\mathbf{S} \multimap \mathbf{S}')$ is defined as follows:

$$id = \{\mathit{copycat}(r) \mid r \text{ is a response of } \mathbf{S}\},$$

where *copycat* is defined as follows:

$$\begin{array}{lcl} \mathit{copycat}(\epsilon) & = & \epsilon \\ \mathit{copycat}(rc) & = & \mathit{copycat}(r)(\mathit{request}\ c)(\mathit{valof}\ c) \\ \mathit{copycat}(qv) & = & \mathit{copycat}(q)(\mathit{is}\ v)(\mathit{output}\ v)\ . \end{array}$$

The word *copycat* used in the description of the identity algorithm has been proposed in [1], and corresponds to a game theoretical understanding: the player always repeats the last move of the opponent. In some influential talks, Lafont had taken images from chess (Karpov – Kasparov) to explain the same thing.

Example 3.2. (1) *The following affine algorithm computes the boolean negation function:*

$$\begin{array}{l} \{(\mathit{request}\ ?)(\mathit{valof}\ ?), \\ (\mathit{request}\ ?)(\mathit{valof}\ ?)(\mathit{is}\ tt)(\mathit{output}\ \mathit{ff}), \\ (\mathit{request}\ ?)(\mathit{valof}\ ?)(\mathit{is}\ \mathit{ff})(\mathit{output}\ tt)\}\ . \end{array}$$

(2) *On the other hand, the left disjunction function cannot be computed by an affine algorithm. Indeed, transcribing the program for* lor *as a strategy leads to:*

$$\begin{array}{l} \{(\mathit{request}\ ?)(\mathit{valof}\ ?.1), \\ (\mathit{request}\ ?)(\mathit{valof}\ ?.1)(\mathit{is}\ tt)(\mathit{output}\ tt), \\ (\mathit{request}\ ?)(\mathit{valof}\ ?.1)(\mathit{is}\ \mathit{ff})(\mathit{valof}\ ?.2), \\ (\mathit{request}\ ?)(\mathit{valof}\ ?.1)(\mathit{is}\ \mathit{ff})(\mathit{valof}\ ?.2)(\mathit{is}\ tt)(\mathit{output}\ tt), \\ (\mathit{request}\ ?)(\mathit{valof}\ ?.1)(\mathit{is}\ \mathit{ff})(\mathit{valof}\ ?.2)(\mathit{is}\ \mathit{ff})(\mathit{output}\ \mathit{ff})\}\ , \end{array}$$

which is not a subset of the set of positions of $Bool^2 \multimap Bool$*, because the projections on* $Bool^2$ *of the last two sequences of moves are not positions of* $Bool^2$*. But the program does transcribe into a (non-affine) sequential algorithm, as we shall see.*

(3) Every constant function gives rise to an affine algorithm, whose responses have the form $(request\, c'_1)(output\, v'_1)\ldots(request\, c'_n)(output\, v'_n)$..

The second and third example above thus justify the terminology affine (in the affine framework, in contrast to the linear one, weakening is allowed). The second example suggests the difference between affine and general sequential algorithms. Both kinds of algorithms ask successive queries to their input, and continue to proceed only after they get responses to these queries. An affine algorithm is moreover required to ask these queries monotonically: each new query must be an extension of the previous one. The 'unit' of resource consumption is thus a sequence of queries/responses that can be arbitrarily large, as long as it builds a position of the input sds. The disjunction algorithms are not affine, because they may have to ask successively the queries ?.1 and ?.2, which are not related by the prefix ordering.

A generic affine algorithm, as represented in figure 1.1, can be viewed as a 'combination' of the following (generic) output strategy and input counter-strategy (or exploration tree):

input counter-strategy | output strategy

$$c \left\{ \begin{array}{l} v_1 \;\cdots \\ \vdots \\ v_i \; d \left\{ \begin{array}{l} \vdots \\ w \\ \vdots \end{array} \right. \\ \vdots \\ v_n \;\cdots \end{array} \right. \qquad\qquad c' \; v' \left\{ \begin{array}{l} c'_1 \;\cdots \\ \vdots \\ c'_m \;\cdots \end{array} \right.$$

We now give a definition of composition of affine algorithms by means of a simple *abstract machine*. Sequential algorithms are syntactic objects, and were indeed turned into a programming language called CDS [7]. What we present here is a simplified version of the operational semantics presented in [14][section 3.5] in the special case of affine algorithms. Given $\phi \in D(\mathbf{S} \multimap \mathbf{S}')$ and $\phi' \in D(\mathbf{S}' \multimap \mathbf{S}'')$, the goal is to compute on demand the positions that belong to their composition ϕ'' in the sds $\mathbf{S} \multimap \mathbf{S}''$. The abstract machine proceeds by rewriting triplets (s, s', s'') where s, s', s'' are positions of $\mathbf{S} \multimap \mathbf{S}'$, $\mathbf{S}' \multimap \mathbf{S}''$, and $\mathbf{S} \multimap \mathbf{S}''$,

$$
\mathit{request}\ c'\ \mathit{valof}\ c
\begin{cases}
\mathit{is}\ v_1\ \cdots \\
\vdots \\
\mathit{is}\ v_i\ \mathit{valof}\ d
\begin{cases}
\vdots \\
\mathit{is}\ w\ \mathit{output}\ v'
\begin{cases}
\mathit{request}\ c'_1\ \cdots \\
\vdots \\
\mathit{request}\ c'_m\ \cdots
\end{cases} \\
\vdots
\end{cases} \\
\vdots \\
\mathit{is}\ v_n\ \cdots
\end{cases}
$$

Figure 1.1. A generic affine algorithm

respectively. The rules are given in Figure 1.2 (where P'' designates the set of positions of $\mathbf{S} \multimap \mathbf{S}''$, etc...):

The first two rules are left to the (streamlike) initiative of the observer. Each time one of these rules is activated, it launches the machine proper, that consists of the four other (deterministic) rules. The generic behaviour of the machine is as follows. The initial triplet is $(\epsilon, \epsilon, \epsilon)$. The observer wants to know the content of c'', or more precisely wants to know what the function does in order to compute the contents of c'' in the output. Thus, he chooses to perform the following rewriting:

$$(\epsilon, \epsilon, \epsilon) \quad \longrightarrow \quad (\epsilon, \epsilon, c'') \ .$$

The request is transmitted to ϕ':

$$(\epsilon, \epsilon, c'') \quad \longrightarrow \quad (\epsilon, c'', c'') \ .$$

There are two cases here. Either ϕ' does not consult its input and produces immediately a value for c'', in which case, this value is transmitted as the final result of the oberver's query:

$$(\epsilon, c'', c'') \quad \longrightarrow \quad (\epsilon, c''v'', c''v'') \qquad (c''v'' \in \phi') \ .$$

Or ϕ' needs to consult its input (like the various sequential or functions), and then an *interaction loop* (in the terminology of Abramsky and Jagadeesan [2]) is initiated:

$$
\begin{array}{rcll}
(\epsilon, c'', c'') & \longrightarrow & (c'_1, c''c'_1, c'') & (c''c'_1 \in \phi') \\
& \longrightarrow & (c'_1v'_1, c''c'_1v'_1, c'') & (c'_1v'_1 \in \phi) \\
& \longrightarrow & (c'_1v'_1c'_2, c''c'_1v'_1c'_2, c'') & (c''c'_1v'_1c'_2 \in \phi') \\
& \vdots & &
\end{array}
$$

$$\begin{array}{llll}
(r, r', r'') & \longrightarrow & (r, r'c'', r''c'') & (r''c'' \in P'') \\
(r, r', r'') & \longrightarrow & (rv, r'c'', r''v) & (r''v \in P'') \\
\\
(r, q', q'') & \longrightarrow & (r, q'v'', q''v'') & (q'v'' \in \phi') \\
(r, q', q'') & \longrightarrow & (rc', q'c', q'') & (q'c' \in \phi') \\
(q, r', q'') & \longrightarrow & (qv', r'v', q'') & (qv' \in \phi) \\
(q, r', q'') & \longrightarrow & (qc, r', q''c) & (qc \in \phi)
\end{array}$$

Figure 1.2. Composition abstract machine for affine algorithms

This dialogue between ϕ and ϕ' may terminate in two ways. Either at some stage ϕ' has received enough information from ϕ to produce a value v'' for c'', i.e. $c'_1v'_1 \ldots c'_nv'_nv'' \in \phi'$:

$$\begin{array}{l}
(c'_1v'_1 \ldots c'_nv'_n, c''c'_1v'_1c'_2 \ldots c'_nv'_n, c'') \\
\quad \longrightarrow (c'_1v'_1 \ldots c'_nv'_n, c''c'_1v'_1c'_2 \ldots c'_nv'_nv'', c''v'')
\end{array}$$

or ϕ itself says it needs to consult its input, i.e., $c'_1v'_1 \ldots c'_nc \in \phi$: this information is passed as a final (with respect to the query c'') result to the observer, who then knows that ϕ'' needs to know the content of c.

$$\begin{array}{l}
(c'_1v'_1 \ldots c'_n, c''c'_1v'_1c'_2 \ldots c'_n, c'') \\
\quad \longrightarrow (c'_1v'_1 \ldots c'_nc, c''c'_1v'_1c'_2 \ldots c'_n, c''c)\ .
\end{array}$$

It is then the observer's freedom to explore further the semantics of ϕ'' by issuing a new query (provided it is in P'') :

$$\begin{array}{l}
(c'_1v'_1 \ldots c'_nv'_n, c''c'_1v'_1c'_2 \ldots c'_nv'_nv'', c''v'') \\
\quad \longrightarrow (c'_1v'_1 \ldots c'_nv'_n, c''c'_1v'_1c'_2 \ldots c'_nv'_nv'', c''v''c''_1)
\end{array}$$

or

$$\begin{array}{l}
(c'_1v'_1 \ldots c'_nc, c''c'_1v'_1c'_2 \ldots c'_n, c''c) \\
\quad \longrightarrow (c'_1v'_1 \ldots c'_nc, c''c'_1v'_1c'_2 \ldots c'_n, c''cv)\ .
\end{array}$$

The query $c''cv$ reads as: "knowing that ϕ'' needs c, how does it behave next when I feed v to c"? After this, the computation starts again using the four deterministic rules along the same general pattern. Notice how ϕ and ϕ' take the leadership in turn (cf. section 1).

We now turn to the abstract definition of our morphisms.

Definition 3.7. A (continuous) function $f : D(\mathbf{S}) \to D(\mathbf{S}')$ is called *stable* if for any $x \in D(\mathbf{S}), \alpha' \in \mathcal{K}(D^\perp$ such that $\mathbf{S}')), f(x) \lhd \alpha'$ there exists a minimum (finite) $y \leq x$ such that $f(y) \lhd \alpha'$ $(m(g, \alpha', x)$, denoted by $m(f, x, \alpha')$. One defines similarly a notion of stable function $g : D^{\perp}(\mathbf{S}') \rightharpoonup D^{\perp}(\mathbf{S})$, with notation $m(g, \alpha', x)$.

Definition 3.8 (symmetric algorithm). Let $\mathbf{S}$ and $\mathbf{S}'$ be two sds's. A symmetric algorithm from $\mathbf{S}$ to $\mathbf{S}'$ is a pair

$$(f : D(\mathbf{S}) \to D(\mathbf{S}'), g : D^{\perp}(\mathbf{S}') \rightharpoonup D^{\perp}(\mathbf{S}))$$

of a function and a partial function that are both continuous and satisfy the following axioms:

$$\text{(L)} \quad (x \in D(\mathbf{S}), \alpha' \in \mathcal{K}(D^{\perp}(\mathbf{S}')), f(x) \lhd \alpha') \Rightarrow \begin{cases} x \lhd g(\alpha') \text{ and} \\ m(f, x, \alpha') = x \lhd\!| \, g(\alpha') \end{cases}$$

$$\text{(R)} \quad (\alpha' \in D^{\perp}(\mathbf{S}'), x \in \mathcal{K}(D(\mathbf{S})), x \rhd g(\alpha') \Rightarrow \begin{cases} f(x) \rhd \alpha' \text{ and} \\ m(g, \alpha', x) = f(x) \, |\!\rhd \alpha' \end{cases}$$

We set as a convention, for any x and any α' such that $g(\alpha')$ is undefined:

$$x \lhd g(\alpha') \text{ and } x \lhd\!| \, g(\alpha') = \emptyset.$$

Hence the conclusion of (L) is simply $m(f, x, \alpha') = \emptyset$ when $g(\alpha')$ is undefined. In contrast, when we write $x \rhd g(\alpha')$ in (R), we assume that $g(\alpha')$ is defined.

Thus, g provides the witnesses of stability of f, and conversely. Moreover, the above definition is powerful enough to imply other key properties of f and g.

Definition 3.9. A (continuous) function $f : D(\mathbf{S}) \to D(\mathbf{S}')$ is called *sequential* if, for any pair $(x, \alpha') \in \mathcal{K}(D(\mathbf{S})) \times \mathcal{K}(D^{\perp}(\mathbf{S}'))$ such that $f(x) \rhd \alpha'$ and $f(z) \lhd \alpha'$ for some $z \geq x$, there exists $\alpha \in \mathcal{K}(D^{\perp}(\mathbf{S}))$, called a *sequentiality index* of f at (x, α'), such that $x \rhd \alpha$ and for any $y \geq x$, $f(y) \lhd \alpha'$ implies $y \lhd \alpha$.

Proposition 3.3. *Let f and g be as in the previous definition. Then f and g are* affine *(i.e., preserve the least upper bounds of pairs of upper bounded elements) and satisfy the following two axioms:*

(LS) If $x \in D(\mathbf{S})$, $\alpha' \in \mathcal{K}(D^{\perp}(\mathbf{S}'))$,$f(x) \rhd \alpha'$, and $f(y) \lhd \alpha'$ for some $y > x$, then $x \rhd g(\alpha')$, and $x \, |\!\rhd g(\alpha')$ is a sequentiality index of f at (x, α').

(RS) If $\alpha' \in D^{\perp}(\mathbf{S}')$, $x \in \mathcal{K}(D(\mathbf{S}))$, $x \lhd g(\alpha')$, and $x \rhd g(\beta')$ for some $\beta' > \alpha'$, then $f(x) \lhd \alpha'$, and $f(x) \lhd\!| \, \alpha'$ is a sequentiality index of g at (α', x). Hence f and g are sequential, and g provides the witnesses of sequentiality for f and conversely.

We turn to the composition of affine algorithms.

Definition 3.10. Let $\mathbf{S}$, $\mathbf{S}'$ and $\mathbf{S}''$ be sds's, and let (f, g) and (f', g') be symmetric algorithms from $\mathbf{S}$ to $\mathbf{S}'$ and from $\mathbf{S}'$ to $\mathbf{S}''$. We define their composition (f'', g'') from $\mathbf{S}$ to $\mathbf{S}''$ as follows:

$$f'' = f' \circ f \quad \text{and} \quad g'' = g \circ g'.$$

The announced full abstraction theorem is the following.

Theorem 3.4. *The sets of affine algorithms and of symmetric algorithms are in a bijective correspondence (actually, an isomorphism), and the two definitions of composition coincide up to the correspondence.*

We just briefly indicate how to pass from one point of view to the other. Given $\phi \in D(\mathbf{S} \multimap \mathbf{S}')$, we define a pair (f, g) of a function and a partial function as follows:

$$\begin{array}{rcl} f(x) & = & \{r' \mid r' = s\lceil_{\mathbf{S}'} \text{ and } s\lceil_{\mathbf{S}} \in x \text{ for some } s \in \phi\} \\ g(\alpha') & = & \{q \mid q = s\lceil_{\mathbf{S}} \text{ and } s\lceil_{\mathbf{S}'} \in \alpha' \text{ for some } s \in \phi\}. \end{array}$$

(By convention, if the right hand side of the definition of g is empty for some α', we interpret this definitional equality as saying that $g(\alpha')$ is undefined.)

Conversely, given a symmetric algorithm (f, g) from $\mathbf{S}$ to $\mathbf{S}'$, we construct an affine algorithm $\phi \in D(\mathbf{S} \multimap \mathbf{S}')$ by building the positions s of ϕ by induction on the length of s (a streamlike process!). For example, if $s \in \phi$, if $s\lceil_{\mathbf{S}}$ and $s\lceil_{\mathbf{S}'}$ are responses, and if $q' = (s\lceil_{\mathbf{S}'})c'$ for some c', then:

$$\begin{array}{ll} sc'c \in \phi & \text{if } (s\lceil_{\mathbf{S}})c \in g(q') \\ sc'v' \in \phi & \text{if } q'v' \in f(s\lceil_{\mathbf{S}}) . \end{array}$$

But, as remarked above, we do not get all sequential functions in this way. Recall that in linear logic the usual implication $A \Rightarrow B$ is decomposed as $(!A) \multimap B$ (!, and its de Morgan dual ?, are called *exponentials* in linear logic).

Definition 3.11 (exponential). Let $\mathbf{S} = (C, V, P)$ be an sds. We set $!\mathbf{S} = (Q, R, P_!)$, where Q and R are the sets of queries and of responses of $\mathbf{S}$, respectively, and where $P_!$ is recursively specified as follows (letting ρ range over responses in $P_!$):

$$\begin{array}{ll} \rho q \in P_! & \text{if } q \in A(\texttt{strategy}(\rho)) \\ & \\ \rho q(qv) \in P_! & \text{if } \rho q \in P_!, \texttt{strategy}(\rho q(qv)) \in D(M), \\ & \text{and } qv \notin \texttt{strategy}(\rho) , \end{array}$$

where `strategy` is the following function mapping responses (or ϵ) of $P_!$ to strategies of $\mathbf{S}$:

$$\texttt{strategy}(\epsilon) = \emptyset \qquad \texttt{strategy}(\rho q(qv)) = \texttt{strategy}(r) \cup \{qv\}.$$

Sequential algorithms between two sds's $\mathbf{S}$ and $\mathbf{S}'$ are by definition affine algorithms between $!\mathbf{S}$ and $\mathbf{S}'$.

It is easily checked that the programs for *lor* (cf. example 3.2), *ror*, *lsor*, and *rsor* transcribe as sequential algorithms from *Bool* $\times$ *Bool* to *Bool*.

Sequential algorithms also enjoy two direct definitions, a concrete one and an abstract one, and both an operational and a denotational definition of composition, for which full abstraction holds, see [14].

Let us end the section with a criticism of the terminology of symmetric algorithm. As already pointed out, the pairs (f, g) are not quite symmetric since g unlike f is a *partial* function. Logically, $\mathbf{S} \multimap \mathbf{S}'$ should read as $\mathbf{S}^\perp \parr \mathbf{S}'$. But something odd is going on: the connective $\parr$ would have two arguments of a different *polarity*: in $\mathbf{S}'$ it is Opponent who starts, while Player starts in $\mathbf{S}^\perp$. For this reason, Laurent proposed to decompose the affine arrow [29] (see also [8]):

$$\mathbf{S} \multimap \mathbf{S}' = (\downarrow \mathbf{S})^\perp \parr \mathbf{S}' ,$$

where $\downarrow$ is a change of polarity operator. For sds's, this operation is easy to define: add a new initial opponent move, call it $\star$, and prefix it to all the positions of $\mathbf{S}^\perp$. For example, $\downarrow (Bool^\perp)$ has $\star\,?\,tt$ and $\star\,?\,f\!f$ as (maximal) positions. According to Laurent's definition, the initial moves of $\mathbf{S_1} \parr \mathbf{S_2}$ are pairs (c_1, c_2) of initial (Opponent's) moves of $\mathbf{S_1}$ and $\mathbf{S_2}$. Then the positions continue as interleavings of a position of $\mathbf{S_1}$ and of $\mathbf{S_2}$. Notice that this is now completely symmetric in $\mathbf{S_1}$ and $\mathbf{S_2}$.

Now, let us revisit the definition of $\mathbf{S} \multimap \mathbf{S}'$. We said that the positions of this sds had to start with a c', which is quite dissymetric. But the $\downarrow$ construction allows us to restore equal status to the two components of the $\parr$. A position in $\mathbf{S}^\perp \parr \mathbf{S}'$ must start with two moves played together in $\mathbf{S}$ and $\mathbf{S}'$. It happens that these moves have necessarily the form $(\star, c')$, which is conveying the same information as c'.

4. Related works

Sequential algorithms turned out to be quite central in the study of sequentiality. First, let us mention that Kleene has developed (for lower types) similar notions [25], under the nice name of oracles, in his late works on the semantics of higher order recursion theory (see [9] for a detailed comparison).

Two important models of functions that have been constructed since turned out to be the extensional collapse (i.e. the hereditary quotient equating sequential algorithms computing the same function, i.e. (in the affine case) two algorithms (f, g) and (f', g') such that $f = f'$): Bucciarelli and Ehrhard's model of strongly stable functions [10, 18], and Longley's model of sequentially realizable functionals [31]. The first model arose from an algebraic characterization of sequential (first-order) functions, that carries over to all types. The second one is a realizability model over a combinatory algebra in which the interaction at work in sequential algorithms is encoded.

Also, Laird has shown that sequential algorithms can be obtained by a collapsing construction from his games model of control in Hyland and Ong style [26].

The models of Hyland and Ong and of Abramsky, Jagadeesan and Malacaria (HO and AJM, respectively) capture PCF definability exactly, whereas the games associated with sequential algorithms also accommodate control operations such as `call-cc` that are not definable in PCF (see section 5). In fact, the interpretation function from normal forms to these models is injective. An essential merit of these works was to characterize the image of this injection, and hence to characterize PCF definability in a syntax-independent way by a few conditions such as innocence and well-bracketing. This opened the way to a whole research program launched by Abramsky. What does happen if one of the conditions is relaxed? Giving up innocence led to very interesting (and fully abstract) models of references (see [4]). Giving up well-bracketing gave a model of PCF plus control, as already mentioned.

The model of sequential algorithms and the HO (or AJM) model differ drastically in size. The type $Bool \to Bool$ is interpreted by a finite sds (i.e., an sds with finitely many positions) in the model of sequential algorithms, while there are infinitely many PCF Böhm trees (and hence infinitely many strategies) in the HO and AJM models at that type. The difference comes from the way the exponential is defined. In definition 3.11, a key feature is non-repetition ($qv \notin \texttt{strategy}(\rho)$). In the games models, the exponential is defined either by interleaving allowing for repetitions or by the opening of potentially infinitely many copies of positions. Roughly, this amounts to dropping the condition $qv \notin \texttt{strategy}(\rho)$.

The finitary nature of sequential algorithms implies that equality in the model is decidable for any type built over *Bool*, while the term model games do not provide effective tools to tackle observational equivalences. As a matter of fact, it has been proved by Loader [30] that equality of two objects in the fully abstract model of (finitary) PCF is undecidable.

A model of PCF is called fully abstract if it equates two terms if and only if these terms are *observationally equivalent*, which means that one can be replaced by the other in any program context without affecting the final result. The *full abstraction problem* of Scott, Milner and Plotkin was the (quite loosely) specified problem of providing a *denotational* construction of the fully abstract model of PCF, as opposed to the original term-based construction of Milner, who also had shown the uniqueness of the fully abstract model [32]. The HO and AJM games models can be called denotational, since they provide a syntax-independent characterization of a term model made of (a PCF version of) Böhm trees. But they yield full abstraction only via a collapse construction which is not essentially different from the one originally performed by Milner. An implicit hope was to arrive at decidability results for the equality in the model, as usual denotational models consist of functions, and hence interpret every type built over Bool by a finite set. Loader's result says that there cannot be such a construction of the fully abstract model of PCF, and justifies a posteriori why game models had to use infinitary exponentials. In contrast, when PCF is extended with control, then the finitary exponential of the model of sequential algorithms does the job (coming next).

5. Control

We already pointed out that theorem 3.4 is a full abstraction result (for the affine case), and that the same theorem has been proved for all sequential algorithms with respect to the language CDS. Sequential algorithms allow inherently to consult the internal behaviour of their arguments and to make decisions according to that behaviour. For example, there exists a sequential algorithm of type $(Bool^2 \to Bool) \to Bool$ that maps *lsor* to *tt* and *rsor* to *ff* (cf. end of section 2). Cartwright and Felleisen made the connection with more standard control operators explicit, and this lead to the full abstraction result of sequential algorithms with respect to an extension of PCF with a control operator [13].

In this respect, the following is a key observation (see [26]). Let o be the sds with one cell and no value: $o = \{\{?\}, \emptyset, \{?\}\}$. Then we have the isomorphism

$$Bool \sim (o \to o \to o) ,$$

where *Bool* is the sds $\{\{?\}, \{tt, ff\}, \{?, (?\ tt), (?\ ff)\}\}$ considered above. Indeed, both sds's have exactly three strategies, ordered in the same way:

$$D(Bool) = \{\emptyset, \{?\ tt\}, \{?\ ff\}\}$$
$$D(o \to o \to o) = \{\emptyset, \{?_\epsilon\ ?_1\}, \{?_\epsilon\ ?_2\}\}$$

(we use subscripts to decorate the cells of the three copies of o, using the convention $o_1 \to o_2 \to o_\epsilon$). It is an instructive exercise to write down explicitly the inverse isomorphisms as sequential algorithms: in one direction, one has the `if then else` function, in the other direction, we have the control operation `catch` considered in [13], which tells apart the two strategies , $\{?_\epsilon ?_1\}, \{?_\epsilon ?_2\}$. Here, we shall show (at type *bool*) how the control operator `call-cc` of Scheme or Standard ML is interpreted as a sequential algorithm of type $((bool \to B) \to bool) \to bool$. The formula $((A \to B) \to A) \to A$ is called Pierce's law and is a typical tautology of classical logic. The connection between control operators and classical logic – and in particular the fact that `call-cc` corresponds to Pierce's law– was first discovered in [21]. Here is is the sequential algorithm interpreting `call-cc` for $A = bool$:

$$?_\epsilon\,?_1 \left\{ \begin{array}{l} ?_{11}\,?_{111} \left\{ \begin{array}{l} tt_{111}\; tt_\epsilon \\ \mathit{ff}_{111}\; \mathit{ff}_\epsilon \end{array} \right. \\ tt_1\; tt_\epsilon \\ \mathit{ff}_1\; \mathit{ff}_\epsilon \end{array} \right.$$

(with labelling of moves $((bool_{111} \to B_{11}) \to bool_1) \to bool_\epsilon$). The same algorithm, with *bool* replaced by $o \to o \to o$, is:

$$?_\epsilon\,?_1 \left\{ \begin{array}{l} ?_{11}\,?_{111} \left\{ \begin{array}{l} ?_{1111}\,?_2 \\ ?_{1112}\,?_3 \end{array} \right. \\ ?_{12}\,?_2 \\ ?_{13}\,?_3 \end{array} \right.$$

(with labelling $(((o_{1111} \to o_{1112}) \to o_{111} \to B_{11}) \to o_{12} \to_{13}\to o_1) \to o_2 \to o_3 \to o_\epsilon$). The reader familiar with continuations may want to compare this tree with the continuation-passing (CPS) style interpretation $\lambda yk.y(\lambda xk'.xk)k$ of `call-cc`, or in tree form (cf. section 1):

$$\lambda yk.y \left\{ \begin{array}{l} \lambda xk'.x \left\{ \; k \right. \\ k \end{array} \right.$$

(where the first k indicates a copy-cat from o_{111} to o_ϵ while the second one indicates a copycat from o_1 to o_ϵ). This is an instance of the injection from terms to strategies mentioned in section 4 (in this simple example, Laird's HO style model coincides with that of sequential algorithms).

CPS translations are the usual indirect way to interpret control operators: first translate, then interpret in your favorite cartesian closed category. In contrast, sequential algorithms look as a direct semantics. The example above suggests that this is an "illusion": once we explicitly replace *bool* by $o \to o \to o$, we find the indirect way underneath.

A more mathematical way to stress this is through thej notion of continuation model [23]: given a category having all the function spaces $A \to R$ for some fixed object R called object of final results, one only retains the full subcategory of *negative* objects, that is, objects of the form $A \to R$. In this category, control can be interpreted. (For the logically inclined reader, notice that thinking of R as the formula "false", then the double negation of A reads as $(A \to R) \to R$, and the classical tautology $((A \to R) \to R) \to A$ is intuitionistically provable for all negative $A = B \to R$.) Now, taking $R = o$, the above isomorphism exhibits *bool* as a negative object. But then all types are negative: given A and $B = B' \to R$, then $A \to B \sim (A \times B') \to R$ is also negative. Hence the model of sequential algorithms (and Laird's model of control) are indeed continuation models, but it is not written on their face.

6. A few more remarks

We would like to mention that this whole line of research on sequential interaction induced such side effects as the design of the Categorical Abstract Machine [11], that gave its name to the language CAML, and of a theory of Abstract Böhm Trees, alluded to in section 1.

As for future lines of research, imports from and into the program of ludics newly proposed by Girard [20] are expected. We just quote one connection with ludics. We insisted in section 2 that *lsor* and *rsor* were different programs for the same function. But there is a way to make them into two different functions, by means of additional *error* values, and accordingly of additional constants in the syntax. Actually, one error is enough, call it *err*. Indeed, we have:

$$lsor(err, \bot) = err \qquad rsor(err, \bot) = \bot\ .$$

Because *lsor* looks at its left argument first, if an error is fed in that argument, it is propagated, whence the result *err*. Because *rsor* looks at its right argument first, if no value is is fed for that argument, then the whole computation is waiting, whence the result $\bot$. One could achieve the same more symmetrically with two different errors: $lsor(err_1, err_2) = err_1$, $rsor(err_1, err_2) = err_2$. But the economy of having just one error is conceptually important, all the more because in view of the isomorphism of section 5, we see that we can dispense (at least for *bool* but also for any finite base type) with the basic values $tt, ff, 0, 1, \ldots$. We arrive then at a picture with only two (base type) constants: $\bot$ and *err*! This is the point of view adopted in Girard's ludics. In ludics, the counterpart of *err* is called Daimon. The motivation for introducing Daimon is quite parallel to that of having errors. Girard's program has the ambition of giving an interactive account of proofs. So, in order to explore a proof of a proposition A, one should play it against a "proof" of $A^\perp$ (the negation of linear logic). But it can't be a proof, since not both A and $A^\perp$ can be proved. So, the space of "proofs" must be enlarged to allow for more opponents to interact with. Similarly, above, we motivated errors by the remark that, once introduced, they allow more observations to be made: here, they allowed us to separate *lsor* and *rsor*. More information, also of a survey kind, can be found in [17].

References

[1] S. Abramsky and R. Jagadeesan, Games and Full Completeness for Multiplicative Linear Logic, Proc. FST-TCS, Springer Lect. Notes in Comp. Sci. 652 (1992).

[2] S. Abramsky and R. Jagadeesan, New foundations for the geometry of iinteraction, Information and Computation 111 (1), 53-119 (1994).

[3] S. Abramsky, R. Jagadeesan, and P. Malacaria, Full abstraction for PCF (1995).

[4] S. Abramsky and G. McCusker, Game semantics, in Computational Logic, U. Berger and H. Schwichtenberg eds, Springer-Verlag, 1-56 (1999).

[5] R. Amadio and P.-L. Curien, Domains and lambda-calculi, Cambridge University Press (1998).

[6] G. Berry and P.-L. Curien, Sequential algorithms on concrete data structures, Theoretical Computer Science 20, 265-321 (1982).

[7] G. Berry and P.-L. Curien, Theory and practice of sequential algorithms: the kernel of the applicative language CDS, in *Algebraic methods in semantics*, Nivat and Reynolds eds., Cambridge University Press, 35-87 (1985).

[8] P. Boudes, PhD thesis (forthcoming).

[9] A. Bucciarelli, Another approach to sequentiality: Kleene's unimonotone functions, Proc. Mathematical Foundations of Programming Semantics, Springer Lect. Notes in Comp. Sci. 802 (1993).

[10] A. Bucciarelli and T. Ehrhard A. Bucciarelli and T. Ehrhard, Sequentiality in an extensional framework, Information and Computation 110(2), 265-296 (1994).

[11] G. Cousineau, P.-L. Curien, and M. Mauny, The categorical abstract machine, Science of Computer Programming 8, 173-202 (1987).

[12] P.-L. Curien, On the symmetry of sequentiality, Proc. Mathematical Foundations of Programming Semantics 1993, Springer Lect. Notes in Comp. Science. 802, 122-130 (1993).

[13] R. Cartwright, P.-L. Curien, and M. Felleisen, Fully abstract semantics for observably sequential languages, Information and Computation 111(2), 297-401 (1994).

[14] P.-L. Curien, Categorical combinators, sequential algorithms and functional programming,Pitman (1986). Revised edition, Birkhäuser (1993).

[15] P.-L. Curien, Abstract Böhm trees, Mathematical Structures in Computer Science 8, 559-591 (1998).

[16] P.-L. Curien and H. Herbelin, Computing with abstract Böhm trees, Third Fuji International Symposium on Functional and Logic Programming, April 1998, Kyoto, Eds Masahiko Sato & Yoshihito Toyama, World Scientific (Singapore), pp 20-39 (1998).

[17] P.-L. Curien, Symmetry and interactivity in programming, to appear in Bulletin of Symbolic Logic.

[18] T. Ehrhard, A relative definability result for strongly stable functions and some corollaries, Information and Computation 152, 111-137 (1999).

[19] J.-Y. Girard, Linear logic, Theoretical Computer Science 50, 1-102 (1987).

[20] J.-Y. Girard, Locus Solum, Mathematical Structures in Computer Science 11(3), 301-506 (2001).

[21] T. Griffin, A formulae-as-types notion of control, Proc. Principles of Programming Languages 1990, ACM Press (1990).

[22] C. Gunter, Semantics of programming languages: structures and techniques, MIT Press (1992).

[23] M. Hofmann and T. Streicher, Continuation models are universal for $\lambda\mu$-calculus, Proc. Logic in Computer Science 1997, Warsaw, IEEE Press (1997).

[24] M. Hyland and L. Ong, On full abstraction for PCF, Information and Computation 163(2), 285-408 (2000).

[25] S. Kleene, Recursive Functionals and Quantifiers of Finite Types Revisited I, II, III, and IV, respectively in Proc. General Recursion Theory II, Fenstad et al. (eds.), North-Holland (1978), Proc. The Kleene Symposium, Barwise et al. (eds.), North-Holland (1980), Proc. Patras Logic Symposium, North Holland (1982), and Proc. Symposia in Pure Mathematics 42 (1985).

[26] J. Laird, Games and sequential algorithms, to appear in Theoretical Computer Science.

[27] F. Lamarche, Sequentiality, games and linear logic, manuscript (1992).

[28] F. Lamarche, Games semantics for full propositional linear logic, Proc. Logic In Computer Science '95, IEEE Computer Society Press (1995).

[29] O. Laurent, Etude de la polarisation en logique, Thèse de Doctorat, Université Aix-Marseille II, march 2002.

[30] R. Loader, Finitary PCF is not decidable, Theoretical Computer Science 266, 341-364 (2001).

[31] J. Longley, The sequentially realizable functionals, to appear in Annals of Pure and Applied Logic.

[32] R. Milner, Fully abstract models of typed lambda-calculi, Theoretical Computer Science 4, 1-23 (1977).

[33] G. Plotkin, LCF as a programming language, Theoretical Computer Science 5, 223-257 (1977).

Chapter 2

UNIVERSAL TYPES AND WHAT THEY ARE GOOD FOR

John R. Longley
School of Informatics
University of Edinburgh
King's Buildings, Edinburgh EH9 3JZ, U.K.
jrl@informatics.ed.ac.uk

Abstract We discuss the standard notions of *universal object* and *universal type*, and illustrate the usefulness of these concepts via several examples from denotational semantics. The purpose of the paper is to provide a gentle introduction to these notions, and to advocate a particular point of view which makes significant use of them. The main ideas here are not new, though our expository slant is somewhat novel, and some of our examples lead to seemingly new results.

Keywords: Category, retract, Karoubi envelope, λ-algebra, universal object, universal type, denotational semantics, extensions of PCF

1. Universal objects

The purpose of this article is to revisit some standard concepts from category theory and denotational semantics: the notions of *universal object* and *universal type*. These ideas are quite simple and are already widely known, but in this article we hope to achieve three things. Firstly, we will try to give a gentle introduction to these ideas for the reader unfamiliar with them. Secondly, although the above concepts themselves are well known, it is perhaps not so widely appreciated how powerful they are — here we wish to advertise these ideas, and to argue that we should perhaps make greater use of them. Finally, we will present some specific applications of the ideas in denotational semantics which may be less familiar, leading to some interesting new results.

Our starting point is the following definition.

GQ Zhang et al. (eds.), Domain Theory, Logic and Computation, 25-63.

Definition 1.1 (Retract). In any category $\mathcal{C}$, an object X is a *retract* of an object Y ($X \lhd Y$) if there are morphisms $f : X \to Y$, $g : Y \to X$ such that $g \circ f = id_X$. We then say the pair (f, g) is a *retraction*.

A helpful intuition is that if X is a retract of Y then the object X can be "coded up" or represented by the object Y. For instance, in concrete categories, we can think of f as coding up elements of X as elements of Y, with g as a corresponding decoding function. More generally, $X \lhd Y$ means that morphisms into X can be coded as morphisms into Y, and likewise morphisms out of X can be coded as morphisms out of Y. Indeed, we have the following easy proposition, which perhaps helps to motivate the definition of a retraction:

Proposition 1.1. *X is a retract of Y iff for any objects A, B there are mappings*

$$\begin{aligned} a \mapsto \overline{a} : &\quad Hom(A, X) \to Hom(A, Y), \\ b \mapsto \tilde{b} : &\quad Hom(X, B) \to Hom(Y, B) \end{aligned}$$

such that for all a, b we have $\tilde{b} \circ \overline{a} = b \circ a$.

The central notion that we shall consider in this paper is the following:

Definition 1.2 (Universal object). In a category $\mathcal{C}$, an object U is *universal* if every object of $\mathcal{C}$ is a retract of U.

If a category $\mathcal{C}$ has a universal object U, the whole of $\mathcal{C}$ can in some sense be coded up in terms of U and its endomorphisms. Firstly, given any object X and a retraction $(f, g) : X \lhd U$, the object X is clearly determined up to isomorphism by the composite $f \circ g : U \to U$. (Notice that this morphism is always an *idempotent*—that is, $(f \circ g)^2 = f \circ g$.) Secondly, given any morphism $h : X \to Y$ and retractions $(f_X, g_X) : X \lhd U$ and $(f_Y, g_Y) : Y \lhd U$, the morphism h is determined by the composite $f_Y \circ h \circ g_X : U \to U$.

Indeed, the whole of $\mathcal{C}$ can essentially be reconstituted from just the monoid M of endomorphisms of U, using the following standard construction. The idea is that the endomorphisms used above to represent objects and morphisms of $\mathcal{C}$ can themselves be taken to be the objects and morphisms of a certain category:

Definition 1.3 (Karoubi envelope). For any monoid $(\mathcal{M}, ., id)$, we may define a category $\mathcal{K}(\mathcal{M})$ (called the *Karoubi envelope* of $\mathcal{M}$) as follows:

- objects are elements $a \in \mathcal{M}$ such that $a.a = a$;

- morphisms $a \to b$ are elements $m \in \mathcal{M}$ such that $b.m.a = m$;
- the identity id_a is simply a;
- given $m : a \to b$ and $n : b \to c$, their composition $n \circ m$ is $n.b.m$.

One can more generally define the Karoubi envelope $\mathcal{K}(\mathcal{D})$ for any category $\mathcal{D}$, but we shall concentrate here on the case of monoids (which can be regarded as one-object categories). Clearly $id \in \mathcal{M}$ is always a universal object of $\mathcal{K}(\mathcal{M})$, and the endomorphisms $id \to id$ are all the elements of $\mathcal{M}$. Moreover, the following fact is easily verified:

Proposition 1.2. *Let $\mathcal{C}$ be a category with a universal object U, and let $\mathcal{M}$ be the monoid of endomorphisms of U. Then there is a full and faithful functor $E : \mathcal{C} \to \mathcal{K}(\mathcal{M})$.*

Thus, the original category $\mathcal{C}$ is equivalent to a full subcategory of the reconstituted category $\mathcal{K}(\mathcal{M})$. Note that $\mathcal{K}(\mathcal{M})$ may be bigger than $\mathcal{C}$, since there may be idempotents $a : U \to U$ which do not arise from retractions in $\mathcal{C}$, but the important thing for our purposes is that we have at least reconstructed all of $\mathcal{C}$— we will not bother too much in this article about the difference between $\mathcal{C}$ and $\mathcal{K}(\mathcal{M})$. (A common situation in denotational semantics is that one has a bunch of categories with some full subcategory in common, and the objects we are really interested in — those corresponding to the types of some programming language — all live in this common subcategory, so we do not care too much about the precise extent of the categories.)

The point of view we want to advocate in this paper is that if $\mathcal{C}$ has a universal object U, and we are able to get a good grasp on the monoid of endomorphisms of U, then we have somehow "cracked" the whole structure of $\mathcal{C}$, since in some sense all the complexity of $\mathcal{C}$ is already present in this monoid. This idea has two main aspects. Firstly, if the relevant monoid $\mathcal{M}$ is itself easy to construct, the definition of $\mathcal{K}(\mathcal{M})$ can offer a simple way of *constructing* the category of interest in the first place—which in at least some cases can be simpler than a more concrete construction of the objects and morphisms of $\mathcal{C}$. Secondly, whether or not we choose to construct our category in this way, once we know it has a universal object, we can often use this to prove *properties* of the category, and in particular to obtain results of interest for denotational semantics.

Of course, this strategy of trying to understand categories of interest via universal objects is not always applicable: the category in question might not have a universal object, or even if it does, its monoid of endomorphisms might not be easy to understand. (We will see examples of both these situations later on.) The point we wish to emphasize, however, is that if there happens to be a readily intelligible universal object, it is certainly worth knowing about it. As we shall see, it turns out that an approach which emphasizes universal objects works well in a large number of interesting cases, and indeed offers a pleasantly "uniform" way of treating a wide variety of situations.

2. λ-algebras

It is natural to ask how particular properties of monoids $\mathcal{M}$ are correlated with properties of the categories $\mathcal{K}(\mathcal{M})$. Here we will consider just one example of such a correlation, namely, the conditions under which $\mathcal{K}(\mathcal{M})$ is a cartesian closed category. The material in this section is fairly standard, and is due mainly to Scott [Sc80] and Koymans [K082]; further details may be found in Barendregt's book [Ba84].

We will assume some basic familiarity with the untyped λ-calculus. To fix notation, for any set A we may consider a set $C(A) = \{c_a | a \in A\}$ of *constant symbols*, one for each element of A. We will write $\Lambda^0(A)$ for the set of closed λ-terms over $C(A)$ (that is, untyped λ-terms with no free variables, possibly involving constants drawn from $C(A)$). We use $M, N, \ldots$ to range over λ-terms, and write $=_\beta$ for the relation of β-equivalence.

The following definition gives one possible notion of a "model" for the untyped λ-calculus:

Definition 2.1 (λ-algebra). A λ-*algebra* is a set A equipped with an "interpretation" function $[\![-]\!] : \Lambda^0(A) \to A$ with the following properties:

$$\begin{array}{rcl} [\![c_a]\!] & = & a, \\ \text{if } M =_\beta N & \text{then} & [\![M]\!] = [\![N]\!]. \end{array}$$

Alternatively (and equivalently), one can define the notion of λ-algebra by means of an interpretation of *open* λ-terms over $C(A)$ (see e.g. [Ba84, Chapter 5]), but this entails a little extra machinery which the above definition avoids. Note that any λ-algebra A can be endowed with a binary operation $\cdot : A \times A \to A$ (called *application*) defined by

$$a \cdot b \;=\; [\![c_a c_b]\!].$$

As an aside, note that every $M \in \Lambda^0(A)$ is β-equivalent to some term built up from constants c_a and the terms $K = \lambda xy.x$, $S = \lambda xyz.xz(yz)$ using just application. It follows that the function $[\![-]\!]$ is completely determined by the above conditions once we have fixed on an application operation and a suitable choice of elements $k = [\![K]\!]$ and $s = [\![S]\!]$.

We will give several examples of λ-algebras in Section 6. In practice, the concrete construction of particular λ-algebras usually conforms to the following general pattern. First, for each finite set F of variables we define a set A_F, elements of which will serve as denotations for λ-terms with free variables drawn from F; we will take A to be $A_\emptyset$. (We may regard elements of A_F as "operations" of type $A^F \to A$, where A^F is the set of all functions from F to A — that is, the set of all valuations for the variables in F. The operations here may be functions of type $A^F \to A$, or they may be algorithms of some intensional kind.) Next we define an interpretation $[\![-]\!]$ on *open* λ-terms by induction on the term structure. It is then usually straightforward to verify that if $M =_\beta N$ then $[\![M]\!] = [\![N]\!]$.

Let us now see how λ-algebras are related to universal objects. Let $\mathcal{C}$ be a cartesian closed category containing an object U such that $U^U \lhd U$. (Such a U is called a *reflexive object*. In a cartesian closed category, any universal object is reflexive.) Then U gives rise to a λ-algebra in the following way. First choose some retraction $U^U \lhd U$; call its components *lambda* : $U^U \to U$ and *apply* : $U \to U^U$. Next, let A be the set of all morphisms $1 \to U$ in $\mathcal{C}$, and let $\cdot : A \times A \to A$ be the function induced by *uncurry*(*apply*) : $U \times U \to U$. Finally, for any term M with free variables drawn from $x_1, \ldots, x_n$, we may define a morphism $[\![M]\!]_\Gamma : U^n \to U$ as follows:

$$\begin{aligned} [\![c_a]\!]_\Gamma &= a \circ !_{U^n} \\ [\![x_i]\!]_\Gamma &= \pi_i \\ [\![MN]\!]_\Gamma &= \mathit{uncurry}(\mathit{apply}) \circ \langle M_\Gamma, N_\Gamma \rangle \\ [\![\lambda x.M]\!]_\Gamma &= \lambda \circ \mathit{curry}([\![M]\!]_\Gamma, x) \end{aligned}$$

(Here, Γ abbreviates $x_1, \ldots, x_n$; $!_X$ is the unique morphism $X \to 1$; π_i is the ith projection $X_1 \times \cdots \times X_n \to X_i$; and *curry*($f$) : $X_1 \times \cdots \times X_i \to Y^{X_{i+1}}$ is the exponential transpose of $f : X_1 \times \cdots \times X_{i+1} \to Y$.) Restricting our attention to closed terms, this yields a function $[\![-]\!] : \Lambda^0(A) \to A$.

Proposition 2.1. *The structure* $(A, \cdot, [\![-]\!])$ *obtained in this way is a* λ-*algebra.*

Proof. The conditions for constants and application are immediate. For the condition regarding β-equivalence, clearly it suffices to check that

$$[\![(\lambda x.M)N]\!]_\Gamma = [\![M[N/x]]\!]_\Gamma$$

for all suitably typed M, N. But this follows easily from the fact that *apply* $\circ$ *lambda* $= id_{U^U}$ (together with the evident connection between syntactic substitution and categorical composition). □

Conversely, given any λ-algebra $(A, \cdot, [\![-]\!])$, we may build a category in which it lives as a reflexive object. Let $(\mathcal{M}_A, ., id)$ be the monoid defined as follows:

- $\mathcal{M}_A = \{a \in A \mid a = [\![\lambda x.c_a x]\!]\} = \{[\![\lambda x.c_a x]\!] \mid a \in A\}$,
- $b.a = [\![\lambda x.c_b(c_a x)]\!]$,
- $id = [\![\lambda x.x]\!]$.

It is easy to check that this is indeed a monoid. Moreover:

Proposition 2.2. *For any λ-algebra A, $\mathcal{K}(\mathcal{M}_A)$ is cartesian closed. If $U = id$, considered as the canonical universal object of $\mathcal{K}(\mathcal{M})$, then U^U is the object given by $[\![\lambda xy.xy]\!] \in \mathcal{M}$. Finally, the λ-algebra arising from the canonical retraction $U^U \lhd U$ is isomorphic to A.*

Proof. For a terminal object in $\mathcal{K}(\mathcal{M})$ we may take $[\![\lambda x.e]\!]$ for any $e \in A$. Given idempotents $a, b \in \mathcal{M}$, the product $a \times b$ is the idempotent

$$[\![\lambda x.pair(c_a(fst\, x))(c_b(snd\, x))]\!],$$

where $pair = \lambda uvz.zuv$, $fst = \lambda uv.u$ and $snd = \lambda uv.v$; and the exponential b^a is $[\![\lambda xy.c_b(x(c_a y))]\!]$. As a special case of the latter, we have $U^U = [\![\lambda xy.xy]\!]$. Finally, morphisms $1 \to U$ in $\mathcal{K}(\mathcal{M})$ are exactly elements of $\mathcal{M}$ of the form $[\![\lambda xy.c_a y]\!]$ for $a \in A$; clearly these correspond precisely to elements of A itself. It is routine to check that the λ-algebra structure arising from U coincides with the original one on A. □

It is also easy to show that if $\mathcal{M}$ is any monoid such that $\mathcal{K}(\mathcal{M})$ is cartesian closed, the resulting λ-algebra gives rise by the above construction to a monoid isomorphic to $\mathcal{M}$. The above results may therefore be summarized as follows:

Theorem 2.3. *Let $\mathcal{M}$ be any monoid. Then $\mathcal{K}(\mathcal{M})$ is cartesian closed iff $\mathcal{M} \cong \mathcal{M}_A$ for some λ-algebra A.*

One might ask whether it is possible to state the condition for $\mathcal{K}(\mathcal{M})$ being cartesian closed purely algebraically in terms of the monoid structure of $\mathcal{M}$. This can indeed be done, but the algebraic conditions are rather more complicated than one might hope for. The details are not especially illuminating and we will not give them here.

The above theorem is just one instance of a correlation between a property of $\mathcal{M}$ and a property of $\mathcal{K}(\mathcal{M})$, and will serve in this paper to illustrate our viewpoint. For the record, we mention a couple of other similar correlations.

First, there is a "linear" analogue of the above correlation: $\mathcal{K}(\mathcal{M})$ is a symmetric monoidal closed category iff $\mathcal{M}$ arises from a linear λ-algebra. (In the linear λ-calculus, we may form the abstraction $\lambda x.M$ only if M contains exactly one free occurrence of x. In fact, we need to work here with a version of the linear λ-calculus which also includes a *pairing* operation $\langle M, N \rangle$ and a *pattern-matching* construct $\lambda\langle x, y\rangle.M$.)

Secondly, if $\mathcal{M}$ arises from a CPO-enriched λ-algebra A (with least element) then $\mathcal{K}(\mathcal{M})$ has initial solutions to many recursive domain equations: essentially, one can solve such an equation by taking the fixed point of a continuous operator on A. This idea was first exploited by Scott in the case of the λ-algebra $\mathcal{P}\omega$ [Sc76]. Correlations of this kind are very useful if we wish to approach the study of particular categories via their universal objects.

3. Denotational semantics

Next we recall some basic concepts from denotational semantics, and explain how universal types may be used to advantage in this setting.

The general situation we consider will consist of the following data. We think of items 1 and 2 below as constituting a programming language, and 3–5 as constituting a denotational model for such a language. The framework given here may look rather cumbersome, but it allows us to state our results in a very general form.

1 A typed language $\mathcal{L}$, consisting of

- a set of types σ,
- an infinite supply of variables $x : \sigma$ for each type σ,
- a set of terms M, such that all variables are terms,
- a *typing relation* $\Gamma \vdash M : \sigma$, where Γ ranges over *environments* $(x_1 : \sigma_1, \ldots, x_n : \sigma_n)$, such that $\Gamma \vdash x : \sigma$ whenever $x : \sigma \in \Gamma$ (we will write $\mathcal{L}^\sigma_\Gamma$ for the set of terms M such that $\Gamma \vdash M : \sigma$),
- an operation of type-respecting *substitution*, yielding for any $\Delta \vdash M : \tau$ (where $\Delta = (x_1 : \sigma_1, \ldots, x_n : \sigma_n)$) and $\Gamma \vdash N_i : \sigma_i$ (for $i = 1, \ldots, n$) a term $\Gamma \vdash M[\vec{N_i}/\vec{x_i}] : \tau$.
- for each Γ, σ and Γ', σ', a set of functions $C(-)$ (which we call *contexts*) mapping terms M such that $\Gamma \vdash M : \sigma$ to terms $C(M)$ such that $\Gamma' \vdash C(M) : \sigma'$. We require, moreover, that contexts constitute the morphisms of a category (whose objects are pairs (Γ, σ)), and that for any well-typed term

$(x : \sigma) \vdash P : \sigma'$ and any environment Γ, the mapping

$$M \mapsto P[M/x] \;:\; \mathcal{L}_\Gamma^\sigma \longrightarrow \mathcal{L}_\Gamma^{\sigma'}$$

is a context.

We call a term M *closed* if $\emptyset \vdash M : \sigma$ for some σ (we usually omit mention of the empty environment).

2 For one or more types σ (called *program types*), a set of special closed terms $V : \sigma$ called *values*, an *evaluation* relation $M \Rightarrow V$ between closed terms and values, and a function Obs_σ mapping values $V : \sigma$ to some set O_σ of *observations.* (We do not require evaluation to be deterministic: we may have $M \Rightarrow V$ and $M \Rightarrow V'$ where V, V' are different values.)

3 A category $\mathcal{C}$ with finite products.

4 An interpretation $[\![-]\!]$ of $\mathcal{L}$ in $\mathcal{C}$, assigning to each type σ an object σ; to each environment $\Gamma = (x_1 : \sigma_1, \ldots, x_n : \sigma_n)$ the object $[\![\Gamma]\!] = [\![\sigma_1]\!] \times \cdots \times [\![\sigma_n]\!]$; and to each typing judgement $\Gamma \vdash M : \sigma$ a morphism $[\![M]\!]_\Gamma : [\![\Gamma]\!] \to [\![\sigma]\!]$. The interpretation is required to satisfy the following conditions:

Variable condition: if $\Gamma = (x_1 : \sigma_1, \ldots, x_n : \sigma_n)$ then $[\![x_i]\!]_\Gamma$ is the ith projection $[\![\Gamma]\!] \to [\![\sigma_i]\!]$.

Compositionality: if $\Gamma \vdash N_i : \sigma_i$ for each i, and $\Delta \vdash M : \tau$ where $\Delta = (x_1 : \sigma_1, \ldots, x_n : \sigma_n)$, then

$$[\![M[\vec{N_i}/\vec{x_i}]]\!]_\Gamma = [\![M]\!]_\Delta \circ \langle [\![N_1]\!]_\Gamma, \ldots, [\![N_n]\!]_\Gamma \rangle : [\![\Gamma]\!] \to [\![\tau]\!].$$

Context condition: if $[\![M]\!]_\Gamma = [\![M']\!]_\Gamma$ and $C(-)$ is a context of appropriate type, then $[\![C(M)]\!]_{\Gamma'} = [\![(C(M')]\!]_{\Gamma'}$.

5 For each program type σ, a relation $\sqsubseteq$ between morphisms $[\![V]\!]$ (V a value of type σ) and morphisms $[\![M]\!]$ (M a closed term of type σ), such that the following requirement is satisfied:

Adequacy: $v \sqsubseteq [\![M]\!]$ iff there is some V such that $[\![V]\!] = v$ and $M \Rightarrow V$.

(In the case of deterministic programming languages, $\sqsubseteq$ will typically be just the equality relation. In non-deterministic settings, $[\![M]\!]$ will be something like a set of possible values for M, and $\sqsubseteq$ will be something like subset inclusion.)

In this situation, we may obtain (from items 1 and 2 alone) a notion of observational equivalence for $\mathcal{L}$:

Definition 3.1 (Observational equivalence). Given $M, N \in \mathcal{L}^\sigma_\Gamma$, we say M, N are *observationally equivalent* at (Γ, σ) ($M \approx^\sigma_\Gamma N$) if for all contexts $C(-) : (\Gamma, \sigma) \to (\emptyset, \tau)$ and values $V : \tau$, we have $C(M) \Rightarrow V$ iff $C(N) \Rightarrow V'$ for some V' with $Obs_\tau(V) = Obs_\tau(V')$.

Remark 3.1. In the case of non-deterministic languages, this notion of observational equivalence is rather a weak one, in at least two respects. Firstly, it does not allow us to distinguish a program that *always* returns 0 from one which may return 0 or may diverge. Secondly, it does not capture anything like *bisimilarity* of terms: for instance, if M, N are of some type such as $unit \to unit \to nat$, then $M \approx N$ will typically tell us only that the possible values of $M(*)(*)$ and $N(*)(*)$ are the same — it does not tell us that for any potential value of $M(*)$ there is a potential value of $N(*)$ with the same possible behaviours. We will not worry too much about these limitations, since non-deterministic languages will play a rather peripheral role in this paper. Nevertheless, it would be interesting to know whether the ideas presented here can be made to work for stronger notions of observational equivalence.

We now introduce two conditions which express "goodness of fit" criteria for a language and a model. Both of these have played major roles in the development of denotational semantics.

Definition 3.2 (Full abstraction, completeness). Let $[\![-]\!]$ be an interpretation for a language $\mathcal{L}$ as above.
(i) $[\![-]\!]$ is *fully abstract* if for all terms $M, N \in \mathcal{L}^\sigma_\Gamma$,

$$M \approx^\sigma_\Gamma N \;\Rightarrow\; [\![M]\!]_\Gamma = [\![N]\!]_\Gamma.$$

(The converse implication holds automatically in our setting.) (ii) $[\![-]\!]$ is *complete* if for every morphism $f : [\![\Gamma]\!] \to [\![\sigma]\!]$ there is a term $M \in \mathcal{L}^\sigma_\Gamma$ with $[\![M]\!]_\Gamma = f$.

The word "universal" is sometimes used here in place of "complete", but in the context of the present paper we prefer not to overload this term too heavily.

It is easy to see that there is essentially only one model of $\mathcal{L}$ that is both fully abstract and complete. More precisely:

Proposition 3.1. *There is exactly one model* $(\mathcal{C}, [\![-]\!])$ *of* $\mathcal{L}$ *(up to equivalence of categories) that is fully abstract and complete, and in which every object of* $\mathcal{C}$ *is a finite product of objects* $[\![\sigma]\!]$. *If* $(\mathcal{D}, [\![-]\!]')$ *is any* *other fully abstract and complete model of* $\mathcal{L}$*, then* $\mathcal{C}$ *is a equivalent to a full subcategory of* $\mathcal{D}$*.*

Proof. We may construct such a model as follows:

- objects of $\mathcal{C}$ are tuples of types $(\sigma_1, \ldots, \sigma_n)$;
- morphisms $(\sigma_1, \ldots, \sigma_n) \rightarrow (\tau_1, \ldots, \tau_m)$ in $\mathcal{C}$ are m-tuples

$$([M_1], \ldots, [M_m]),$$

  where $(x_1 : \sigma_1, \ldots, x_n : \sigma_n) \vdash M_i : \tau_i$, and $[M_i]$ is the observational equivalence class of M_i;
- the interpretation $[\![-]\!]$ of $\mathcal{L}$ in $\mathcal{C}$ is the canonical one;
- the relation $\sqsubseteq$ is given by $v \sqsubseteq f$ iff $v = [\![V]\!]$ and $f = [\![M]\!]$ for some $V, M : \sigma$ where $M \Rightarrow V$.

Given any other such model $\mathcal{C}'$, we may assume the objects of $\mathcal{C}'$ are tuples $(\sigma_1, \ldots, \sigma_n)$, and by completeness, the morphisms $(\sigma_1, \ldots, \sigma_n) \rightarrow (\tau_1, \ldots, \tau_m)$ must be tuples of terms M_i as above modulo some equivalence relation $\sim$. By full abstraction, $\sim$ must include componentwise observational equivalence; but by the context condition and adequacy, $\sim$ cannot be any larger than this. Thus $\mathcal{C}'$ is equivalent to $\mathcal{C}$. The second statement of the Proposition follows immediately from the first. □

The model $\mathcal{C}$ in this proposition is in some sense the canonical or "best possible" model of $\mathcal{L}$. The above proof shows how $\mathcal{C}$ can be obtained as the syntactic model (or *term model*) of $\mathcal{L}$; however, this way of constructing a model is not very enlightening, since proving things about this model is no easier than proving things directly about $\mathcal{L}$ itself. In general, we would prefer a more "semantic" characterization of $\mathcal{C}$ which gave us a better mathematical handle on its structure — in other words, which offered us some understanding of $\mathcal{C}$ that was independent of the syntax of $\mathcal{L}$. Indeed, one possible working philosophy for denotational semantics (which the present author finds appealing) is that its goal should be to give good mathematical characterizations of the fully abstract and complete model for various languages $\mathcal{L}$.

Note in passing that since most programming languages of interest are executable in an "effective" fashion, a universal model will inevitably need some notion of effectivity built in, to ensure that all morphisms are in some sense computable. Usually we do this simply by making use of the set of partial recursive functions at some point in the construction.

4. Universal types

We now show how the idea of universal objects can be of use in denotational semantics. Given a model $(\mathcal{C}, [\![-]\!])$ of a language $\mathcal{L}$ as in items 1–5 above, we may introduce the following notion:

Definition 4.1 (Universal type). (i) A type ρ is a *definable retract* of σ with respect to $[\![-]\!]$ if there are well-typed terms

$$x : \rho \vdash F : \sigma, \qquad y : \sigma \vdash G : \rho$$

such that $[\![G]\!]_{(x:\rho)} \circ [\![F]\!]_{(y:\sigma)} = id_{[\![\rho]\!]}$.

(ii) A type σ is a *universal type* for $[\![-]\!]$ if every type of $\mathcal{L}$ is a definable retract of σ w.r.t. $[\![-]\!]$.

Clearly, if σ is a universal type and σ is a definable retract of τ then τ is also a universal type. Thus, if there exists a universal type there are likely to be several. Usually, of course, we are interested in identifying the simplest possible universal type.

Of course, it frequently happens that σ is a universal type for $[\![-]\!]$ *and* $[\![\sigma]\!]$ is a universal object in $\mathcal{C}$, although neither fact implies the other in general: there may be non-denotable objects of $\mathcal{C}$ that are not retracts of $[\![\sigma]\!]$, and conversely the components of a retraction $[\![\rho]\!] \lhd [\![\sigma]\!]$ need not be definable in $\mathcal{L}$. It can also happen that $\mathcal{C}$ has a universal object, but none of the form $[\![\sigma]\!]$.

If the types of $\mathcal{L}$ are inductively generated in some way, one can often prove by induction on the types that some type σ is universal. For example:

Proposition 4.1. *Suppose $\mathcal{L}$ is a simply typed λ-calculus over some set of* ground *types γ — that is, the types of $\mathcal{L}$ are given by the grammar $\sigma ::= \gamma \mid \sigma_1 \to \sigma_2$. Suppose also that $M =_{\beta\eta} M'$ implies $[\![M]\!]_\Gamma = [\![M']\!]_\Gamma$ for any suitable Γ. Then σ is universal for $[\![-]\!]$ iff all the ground types plus the type $\sigma \to \sigma$ are definable retracts of σ.*

Proof. The left-to-right implication is immediate. For the converse, we show by induction that all types ρ are definable retracts of σ. If ρ is a ground type this is true by hypothesis. If $\rho = \rho_1 \to \rho_2$ where the ρ_i are definable retracts of σ via terms

$$x_i : \rho_i \vdash F_i : \sigma, \qquad y_i : \sigma \vdash G_i : \rho_i$$

$(i = 1, 2)$, then ρ is a definable retract of $\sigma \to \sigma$ via the terms

$$\begin{aligned} f : \rho \quad &\vdash (\lambda y_1 : \sigma . F_2[f(G_1)/x_2]) : \sigma \to \sigma, \\ g : \sigma \to \sigma \quad &\vdash (\lambda x_1 : \rho_1 . G_2[g(F_1)/y_2]) : \rho \end{aligned}$$

But $\sigma \to \sigma$ is itself a definable retract of σ, so the result follows easily. □

For some languages we wish to consider (particularly call-by-value languages), full $\beta\eta$-equality does not hold in the natural models. For such languages, a slightly more refined (but more cumbersome) version of the above proposition is useful:

Proposition 4.2. *Suppose $\mathcal{L}$ is a simply typed λ-calculus over a set of ground types, in which certain closed terms (including all of the form $\lambda x.M$) are designated as values. Suppose also that $[\![-]\!]$ is an interpretation such that*

- *if V is a value then $[\![(\lambda x.M)V]\!]_\Gamma = [\![M[V/x]]\!]_\Gamma$ for any suitable Γ;*
- *for any type $\sigma = \sigma_1 \to \sigma_2$, $[\![(\lambda fx.fx)f]\!]_{(f:\sigma)} = [\![f]\!]_{(f:\sigma)}$.*

Let α be any type for which there exists some value $U : \alpha$. Then σ is universal for $[\![-]\!]$ iff all ground types plus the type $(\alpha \to \sigma) \to \sigma$ are definable retracts of σ.

Proof. As in the previous proposition, except that if $\rho = \rho_1 \to \rho_2$ we use the retraction $[\![\rho]\!] \lhd [\![(\alpha \to \sigma) \to \sigma]\!]$ defined by the terms

$$\begin{array}{rl} f : \rho_1 \to \rho_2 & \vdash (\lambda y_1 : \alpha \to \sigma.F_2[f(G_1[y_1 U])]) : (\alpha \to \sigma) \to \sigma, \\ g : (\alpha \to \sigma) \to \sigma & \vdash (\lambda x_1 : \rho_1.G_2[g(\lambda z : \alpha.F_1[x_1])]) : \rho. \end{array}$$

□

The examples we shall focus on in this paper will all be of the above kind. However, the same principle clearly applies to many languages with other type constructors.

We come at last to the connection between universal types and the key ideas of denotational semantics. The following observation, though trivial to prove, turns out to be very powerful:

Proposition 4.3. *Suppose σ is a universal type for $[\![-]\!]$. Then*
(i) $[\![-]\!]$ is fully abstract iff it is fully abstract for terms

$$M \in \mathcal{L}^{\sigma}_{(x_1:\sigma,\ldots,x_n:\sigma)}$$

in the obvious sense.
(ii) $[\![-]\!]$ is complete iff it is complete for morphisms

$$f : [\![\sigma]\!] \times \cdots \times [\![\sigma]\!] \to [\![\sigma]\!].$$

In the case of typed λ-calculi, we can simplify this further:

Proposition 4.4. *Suppose $\mathcal{L}$ is a typed λ-calculus, $[\![-]\!]$ respects β-equality as in Proposition 4.1, and σ is a universal type for $[\![-]\!]$. Then*
(i) $[\![-]\!]$ is fully abstract iff it is fully abstract for closed terms $M : \sigma$.
(ii) $[\![-]\!]$ is complete iff it is complete for morphisms $1 \to [\![\sigma]\!]$.

These results provide a useful tool for showing that models are fully abstract or complete. As we shall see in the next section, this often leads to simpler proofs than those obtained by more traditional methods.

Definition 4.1 above gives a notion of universal type relative to a particular interpretation $[\![-]\!]$. However, we may also define the following notion, which depends only on $\mathcal{L}$ and its evaluation relation:

Definition 4.2 (Universal type for a language). A type σ is a *universal type for $\mathcal{L}$* iff it is a universal type for the fully abstract and complete model of $\mathcal{L}$ — that is, if for all ρ there are terms $x : \rho \vdash F : \sigma$, $y : \sigma \vdash G : \rho$ such that $G[F/y] \approx x$.

The following proposition is clear:

Proposition 4.5. *Suppose $(\mathcal{C}, [\![-]\!])$ is the term model of $\mathcal{L}$. Then $[\![\sigma]\!] = (\sigma)$ is a universal object in $\mathcal{C}$ iff σ is a universal type for $\mathcal{L}$ and $(\sigma, \sigma) \lhd (\sigma)$.*

(Note that the condition $(\sigma, \sigma) \lhd (\sigma)$ holds automatically under various mild conditions: for instance, if $\mathcal{L}$ has well-behaved product types, or if $\mathcal{L}$ is a typed λ-calculus admitting a representation of the booleans.)

In the next section we will see some examples of universal types for particular languages. In a sense, a universal type is a rabbit pulled from a hat — there is no obvious general method for finding a universal type for a language if there is one — but once we have pulled this rabbit, we can often use it to good effect. For example, in the case of simply typed λ-calculi, the following general line of attack suggests itself. First, give a semantic construction of a category $\mathcal{C}$, either by constructing an appropriate λ-algebra and taking its Karoubi envelope, or otherwise. Next, give a simply typed λ-calculus $\mathcal{L}$, an interpretation $[\![-]\!]$ of $\mathcal{L}$ in $\mathcal{C}$, and a type σ of $\mathcal{L}$, such that:

1 $[\![-]\!]$ validates $\beta\eta$-equality [resp. the conditions of Proposition 4.2],

2 all ground types are definable retracts of σ with respect to $[\![-]\!]$,

3 $\sigma \to \sigma$ [resp. $(\alpha \to \sigma) \to \sigma$] is a definable retract of σ w.r.t. $[\![-]\!]$,

4 $[\![-]\!]$ is fully abstract for closed terms $M : \sigma$,

5 $[\![-]\!]$ is complete for morphisms $1 \to U$.

In practice these are usually trivial to verify except for fact 3, which typically just requires a little programming in $\mathcal{L}$ to define a suitable retraction. It then follows from the above results that $(\mathcal{C}, [\![-]\!])$ is a fully abstract and complete model for $\mathcal{L}$.

Of course, this does not give us much of a clue how to construct a good semantic model for a given language, though once we have found one (together with a universal type) it often offers (perhaps with hindsight) a smooth route to the relevant results.

5. Syntax and semantics of PCF

Next, we wish to consider how the above ideas work in practice for a variety of particular languages. For convenience, we will formulate all our languages as extensions of PCF, the prototype functional programming language first studied by Scott [Sc93] and Plotkin [Pl77]. PCF embodies an appealing notion of sequential, functional computation at higher types, and in essence forms the basis of programming languages such as Standard ML and Haskell. From our point of view, PCF provides a convenient "core language" to which we may add various other computational features in order to investigate the expressive power and the semantics of the resulting languages. In this section we will review the information we will need concerning the syntax and semantics of PCF.

In fact, it will be convenient to define two slightly different versions of PCF — the original *call-by-name* version PCF_N as in [Pl77], and a *call-by-value* version PCF_V as in e.g. [Si90]. From our point of view, the difference between the two versions is not very significant, and all the results we mention will transfer easily from one to the other. However, for several of the extensions of PCF that we consider, one version of the language turns out to be more convenient. (This is because, in the categories we consider, the simplest and most natural universal object corresponds sometimes to a call-by-name type, and sometimes to a call-by-value type.)

The syntax of the two versions is almost identical. We take PCF_N to be the simply typed λ-calculus with ground types *nat*, *bool* and *unit*, together with constants

$$\begin{array}{rlrl} * & : unit, & \mathsf{true},\ \mathsf{false} & : bool, \\ 0,1,2,\ldots & : nat, & \mathsf{succ},\ \mathsf{pred} & : nat \to nat, \\ \mathsf{iszero} & : nat \to bool, & \mathsf{cond}_\gamma & : bool \to \gamma \to \gamma \to \gamma, \\ \mathsf{Y}_\sigma & : (\sigma \to \sigma) \to \sigma & & \end{array}$$

where γ ranges over ground types and σ over all types. The definition for PCF_V is the same except that σ ranges only over *arrow* types.

The operational semantics for PCF_N is given as follows. We declare the ground types (only) to be program types, and designate as *values* V the terms $*$, true, false, $0, 1, 2, \ldots$. For each ground type γ, we take O_γ to be the set of values of type γ and Obs_γ to be the identity. The evaluation relation $M \Rightarrow V$ is defined via the following derivation rules. (Note that $\smile$ denotes truncated subtraction: $m \smile n = \max(m-n, 0)$. We also define $iszero(0) = \mathsf{true}$ and $iszero(n+1) = \mathsf{false}$.)

$$\frac{}{V \Rightarrow V} \qquad \frac{M \Rightarrow n}{\mathsf{succ}\ M \Rightarrow n+1}$$

$$\frac{M \Rightarrow n}{\mathsf{pred}\ M \Rightarrow n \smile 1} \qquad \frac{M \Rightarrow n}{\mathsf{iszero}\ M \Rightarrow iszero(n)}$$

$$\frac{M \Rightarrow \mathsf{true} \quad N \Rightarrow V}{\mathsf{cond}_\gamma\ MNP \Rightarrow V} \qquad \frac{M \Rightarrow \mathsf{false} \quad P \Rightarrow V}{\mathsf{cond}_\gamma\ MNP \Rightarrow V}$$

$$\frac{M[N/x]P_1 \ldots P_r \Rightarrow V}{(\lambda x.M)NP_1 \ldots P_r \Rightarrow V} \qquad \frac{M(\mathsf{Y}_\sigma M)P_1 \ldots P_r \Rightarrow V}{\mathsf{Y}_\sigma MP_1 \ldots P_r \Rightarrow V}$$

The operational semantics of PCF_V is somewhat more complicated. Here we declare *all* types to be program types. We designate as values V all the ground type values as before, together with all closed terms of the form $\lambda x.M$. For ground types γ, we take O_γ and Obs_γ as before; for arrow types σ, we take O_σ to be the one-element set and Obs_σ the unique mapping into this set. (Intuitively, ground type values are observable, but at higher types only the fact of termination is observable.) The evaluation relation is defined by the same rules as above, except that in place of the last two rules we have

$$\frac{M \Rightarrow \lambda x.M' \quad N \Rightarrow V \quad M'[V/x] \Rightarrow V'}{MN \Rightarrow V'} \qquad \frac{}{\mathsf{Y}_\sigma M \Rightarrow \lambda x.M(\mathsf{Y}_\sigma M)x}$$

The first of these rules encapsulates the fundamental idea of call-by-value: a term N must be *evaluated* before it can be passed as a parameter to a function. The absence of the additional arguments $P_1 \ldots P_r$ in both these rules corresponds to the fact that we have values at all types, not just ground types.

Several minor variations in these definitions are possible — for instance, we could dispense with the types *bool* and *unit*, or add product types — but these variations are not very interesting from the point of view of denotational semantics since all the variants are easily intertranslatable. In fact, even PCF_N and PCF_V admit good translations into each other and are in some sense just different presentations of the same thing (see e.g. [Lo95, Chapter 6]). Similar remarks apply to the extensions we consider. (We will return to this point in Section 7.) In the examples to follow, the choice between the two versions is thus more a matter of convenience and taste than of mathematical substance.

We will view the above languages, and the extensions to be introduced, as instances of the scenario described in Section 3. In all cases, we will take the notion of substitution to be the usual operation of simultaneous capture-avoiding substitution, and the notion of context to be given by the usual notion of term context (admitting textual substitutions which may result in variable capture).

We will also consider interpretations of a certain standard kind in cartesian closed categories. Often we can use the same cartesian closed category $\mathcal{C}$ for both the call-by-name and call-by-value versions of a language (in many cases, $\mathcal{C}$ will arise as the Karoubi envelope of some λ-algebra). Let us consider how to model a call-by-name language $\mathcal{L}_N$ extending PCF_N. First, for each ground type γ of $\mathcal{L}_N$ we choose an object X_γ of $\mathcal{C}$. Unless otherwise stated, the objects $X_{nat}, X_{bool}, X_{unit}$ will be the evident objects $\mathrm{N}_\perp, 2_\perp, 1_\perp$ respectively. We can then interpret the types of $\mathcal{L}_N$ as follows:

$$[\![\gamma]\!] = X_\gamma, \qquad [\![\sigma_1 \to \sigma_2]\!] = [\![\sigma_2]\!]^{[\![\sigma_1]\!]}$$

Next, for each constant $c : \sigma$ of $\mathcal{L}_N$ we specify a suitable morphism $[\![c]\!] : 1 \to [\![\sigma]\!]$. The intended interpretation of the constants is usually obvious. For example, we generally take $[\![\mathsf{Y}_\sigma]\!]$ to be the exponential transpose of some fixed point operator $\mathit{Fix}_\sigma : [\![\sigma]\!]^{[\![\sigma]\!]} \to [\![\sigma]\!]$; in the case that $\mathcal{C} \equiv \mathcal{K}(A)$, such operators arise in a standard way from a single fixed point operator on A itself. We may then extend the interpretation to all terms by means of the clauses:

$$\begin{aligned} [\![x_i]\!]^\Gamma &= \pi_i \ \ (\text{where } \Gamma = (x_1 : \sigma_1, \ldots, x_n : \sigma_n)), \\ [\![\lambda x : \sigma . M]\!]^\Gamma &= \mathit{curry}([\![M]\!]^{\Gamma, x:\sigma}), \\ [\![MN]\!]^\Gamma &= \mathit{eval} \circ \langle [\![M]\!]^\Gamma, [\![N]\!]^\Gamma \rangle. \end{aligned}$$

where *curry* and *eval* are given by the cartesian closed structure of $\mathcal{C}$. Finally, unless otherwise stated, we take the relation $\sqsubseteq$ to be simply the identity relation on morphisms of the form $[\![V]\!]$.

The interpretation of call-by-value languages is slightly more involved. Here we require additionally that $\mathcal{C}$ has ω-fold products, and is equipped with a "lifting" monad $(\perp, \eta, \mu)$. Suppose $\mathcal{L}_V$ is an extension of PCF_V. As before, we take an object X_γ for each ground type γ; in all the examples we shall consider, the choice of X_γ will be standard. We now define for each type σ an object $[\![\sigma]\!]$, and for each *arrow* type σ an object $[\![\sigma]\!]$, by mutual recursion as follows:

$$\begin{array}{llll} [\![\gamma]\!] & = X_\gamma, & [\![\sigma_1 \to \sigma_2]\!] & = \lceil\!\lceil \sigma_1 \to \sigma_2 \rceil\!\rceil_\bot, \\ \lceil\!\lceil \gamma \to \sigma_2 \rceil\!\rceil & = [\![\sigma_2]\!]^{|\gamma|}, & \lceil\!\lceil \sigma_1 \to \sigma_2 \rceil\!\rceil & = [\![\sigma_2]\!]^{\lceil\!\lceil \sigma_1 \rceil\!\rceil} \qquad (\sigma_1 = \tau \to \tau'). \end{array}$$

Here, by definition, $|unit| = 1$, $|bool| = 2$, and $|nat| = \omega$; the notation $X^{|\gamma|}$ simply means the categorical product in $\mathcal{C}$ of $|\gamma|$ copies of X. The only reason for the distinction between ground types and arrow types in the above definition is that $\mathcal{C}$ may lack suitable objects to play the role of $\lceil\!\lceil \gamma \rceil\!\rceil$.

Next, one again specifies a morphism $[\![c]\!] : 1 \to [\![\sigma]\!]$ for each constant c. One then extends the interpretation to all terms by means of clauses for variables, abstraction and application. The following works for all the call-by-value examples treated below:

$$\begin{array}{ll} [\![x_i]\!]^\Gamma & = \pi_i, \\ [\![\lambda x : \sigma.M]\!]^\Gamma & = \eta \circ [\![\tau]\!]^\eta \circ curry([\![M]\!]^{\Gamma, x:\sigma}), \\ [\![MN]\!]^\Gamma & = eval \circ \langle strict \circ \alpha_{\sigma\to\tau} \circ [\![M]\!]^\Gamma, [\![N]\!]^\Gamma \rangle. \end{array}$$

(In the second clause, we assume M has type τ, and if σ is a ground type γ then we understand $[\![\tau]\!]^\eta$ to mean the evident map $[\![\tau]\!]^{[\![\gamma]\!]} \to [\![\tau]\!]^{|\gamma|}$. In the third clause, $\alpha_{\sigma\to\tau}$ is the evident map $\lceil\!\lceil \sigma \to \tau \rceil\!\rceil_\bot \to \lceil\!\lceil \sigma \to \tau \rceil\!\rceil$, and *strict* : $B_\bot^A \to B_\bot^{A_\bot}$ is the morphism that extends a map $A \to B_\bot$ to a strict map $A_\bot \to B_\bot$.) As before, the relation $\sqsubseteq$ is the identity on morphisms $[\![V]\!]$ unless otherwise stated.

For all the languages $\mathcal{L}_N, \mathcal{L}_V$ and all the interpretations we shall consider, the proof of adequacy is completely routine (the standard method of proof as in [Pl77] works without any problems).

6. Examples of universal types

We will now investigate how the ideas of Sections 1–4 work out for a variety of extensions of PCF. Since our main purpose is to survey a large number of instances of the generic situation outlined above, we will not present all the examples in full detail but will refer freely to other papers, concentrating here on those points that are not covered elsewhere in the literature.

6.1. PCF itself

We start with a negative result: for the language PCF itself (either PCF_N or PCF_V) there is no universal type. This can be proved fairly easily using the fully abstract and complete game models described e.g.

in [HO00, AM99]. (The result is folklore among the experts in game semantics, but as far as we know a proof has not yet appeared in the literature.) It would seem, intuitively, that the term model of PCF increases indefinitely in its complexity as one goes up the types.

Of course, one can conservatively extend PCF by adding recursive types (the resulting language is known as FPC) — it is then fairly easy to show that a recursively defined type such as $\mu\sigma.nat + (\sigma \to \sigma)$ is universal. However, there would seem to be little gain in doing this, since we do not have any particularly good handle on the structure of this type and its monoid of endomorphisms. Indeed, it would seem that in the known fully abstract and complete models for FPC, it is really no easier to understand this type than to understand the finite approximations from which it is constructed, viz.

$$\sigma_0 = \emptyset, \qquad \sigma_{n+1} = nat + (\sigma_n \to \sigma_n),$$

which is tantamount to understanding all the PCF types in the first place. It would therefore seem that our approach via universal types has little to offer in the case of pure PCF.

6.2. PCF with parallel functions

We now consider the well known extension of PCF with "parallel" computable functionals discovered by Plotkin [Pl77] and Sazonov [Sa76]. Here it suits our purposes to work with the call-by-value version. Specifically, let PCF_V^{++} be the language obtained by extending the definition of PCF_V with constants

$$\begin{array}{rl} \mathsf{por} & : (unit \to bool) \to (unit \to bool) \to bool, \\ \mathsf{exists} & : ((unit \to nat) \to bool) \to bool \end{array}$$

(where por abbreviates "parallel or"), and evaluation rules

$$\frac{M \Rightarrow \mathsf{true}}{\mathsf{por}\ MN \Rightarrow \mathsf{true}} \qquad \frac{N \Rightarrow \mathsf{true}}{\mathsf{por}\ MN \Rightarrow \mathsf{true}} \qquad \frac{M \Rightarrow \mathsf{false} \quad N \Rightarrow \mathsf{false}}{\mathsf{por}\ MN \Rightarrow \mathsf{false}}$$

$$\frac{MP_n \Rightarrow \mathsf{true}}{\mathsf{exists}\ M \Rightarrow \mathsf{true}} \qquad \frac{MP_\bot \Rightarrow \mathsf{false}}{\mathsf{exists}\ M \Rightarrow \mathsf{false}}$$

where $P_n = \lambda x.n$ for $n = 0, 1, \ldots$, and $P_\bot = \mathsf{Y}(\lambda t.t)$.

It is well known that a fully abstract and complete model for PCF^{++} is provided by the category of effective Scott domains and computable maps (see [Pl77]), or indeed by its full subcategory $\mathbf{Coh}_{\mathit{eff}}$ of *effective coherent domains.* It is also known that the object $\mathbb{T}^\omega = 2_\bot^{\mathbb{N}}$ is a universal object in this category, and indeed that $\mathbf{Coh}_{\mathit{eff}}$ is equivalent to the

Karoubi envelope of the corresponding λ-algebra $\mathbb{T}^{\omega}_{\mathit{eff}}$ (see [Pl78]). Since $\mathbb{T}^{\omega} = [\![\, nat \to bool]\!]$, it follows that $nat \to bool$ is a universal type for PCF^{++}_{V}. (Note that $nat \to bool$ is also a universal type for the call-by-name analogue PCF^{++}_{N}, but that in this case the corresponding object of $\mathbf{Coh}_{\mathit{eff}}$ is the more complicated object $2_{\perp}^{\mathrm{N}_{\perp}}$.)

Let us now use this universal type to reconstruct the above results. It is possible to define the λ-algebra $\mathbb{T}^{\omega}_{\mathit{eff}}$ concretely, and then obtain the category $\mathbf{Coh}_{\mathit{eff}}$ as its Karoubi envelope (the details are more or less covered in [Pl78]), though this seems slightly perverse as it is probably easier to construct $\mathbf{Coh}_{\mathit{eff}}$ directly. In any case, we may then give an adequate interpretation $[\![-]\!]$ of PCF^{++} in $\mathbf{Coh}_{\mathit{eff}}$ as usual. Let σ be the type *nat* $\to$ *bool*. Regarding the five conditions given at the end of Section 4, it is clear that $[\![-]\!]$ validates the conditions mentioned in Proposition 4.2, and that *nat*, *bool* and *unit* are definable retracts of σ. It is also trivial that $[\![-]\!]$ is fully abstract and complete for the type σ: just note that terms denoting different functions are observationally distinguishable, and that PCF^{++} is complete for first-order partial recursive functions. To show full abstraction and completeness at all types, It therefore only remains to supply terms

$$f : (unit \to \sigma) \to \sigma \vdash \mathit{Lambda}[f] : \sigma,$$
$$x : \sigma \vdash \mathit{Apply}[x] : (unit \to \sigma) \to \sigma$$

that define a retraction $[\![\,(unit \to \sigma) \to \sigma\,]\!] \lhd [\![\,\sigma\,]\!]$. It is an interesting (and somewhat non-trivial) exercise in programming in PCF^{++} to find suitable terms which correspond to the retraction $[\mathbb{T}^{\omega} \to \mathbb{T}^{\omega}] \lhd \mathbb{T}^{\omega}$ described semantically in [Pl78]. However, the terms themselves are slightly grungy and we omit them here.

Finally, we have to check that the terms *do* define a retraction. If $\mathbf{Coh}_{\mathit{eff}}$ has been defined concretely, this amounts to showing that the interpretation of

$$f : (unit \to \sigma) \to \sigma \vdash \mathit{Apply}[\mathit{Lambda}[f]]$$

is the identity map on $[\![\,(unit \to \sigma) \to \sigma\,]\!]$. If $\mathbf{Coh}_{\mathit{eff}}$ has been obtained as the Karoubi envelope of $\mathbb{T}^{\omega}_{\mathit{eff}}$, it amounts to showing that the interpretation of

$$x : \sigma \vdash \mathit{Lambda}[\mathit{Apply}[x]]$$

coincides with the idempotent on $\mathbb{T}^{\omega}_{\mathit{eff}}$ corresponding to the untyped λ-term $\lambda fx.fx$. In either case, the verification is straightforward but tedious, owing to the complexity of the terms *Lambda* and *Apply*.

It is a moot point whether this is any simpler than the usual proof of full abstraction and completeness (given in [Pl77]), but it at least offers an interesting alternative. In particular, it is distinguished from the usual proof by the fact that only relatively low types are involved.

Another advantage of our proof is that it is easily *portable*: once the terms *Lambda* and *Apply* have been constructed, they can be re-used for other models of PCF^{++}. For example, we can now show cheaply that the interpretation of PCF^{++} in the category of PERs over the natural numbers (see e.g. [Lo95]) is fully abstract and complete. Here again, all the other conditions are trivial to verify, so we just need to show that the above terms do define a retraction in this model. But this follows immediately from the classical Myhill-Shepherdson theorem, which tells us that the elements of $[\![(unit \to \sigma) \to \sigma]\!]$ in this category are the same as in $\mathbf{Coh}_{\mathit{eff}}$.

6.3. PCF with non-determinism

Next, let us consider an extension of PCF with non-deterministic choice. In this case, the call-by-name version is definitely simpler. Specifically, we add to the definition of PCF_N a single new constant

$$\mathsf{choose} : nat$$

together with the evaluation rule

$$\frac{}{\mathsf{choose} \Rightarrow n}$$

where n ranges over all numerals. To give an adequate semantics for this language, the interpretation of the type *nat* cannot be just the usual set $N_\perp$, but will need to incorporate all r.e. subsets of *nat*; the denotation of a closed term $M : nat$ will then be the set of all n such that $M \Rightarrow n$. This suggests that we work with Scott's λ-algebra $\mathcal{P}\omega$, or rather its effective submodel $\mathcal{P}\omega_{re}$, whose underlying set is the set of r.e. subsets of N (see [Sc76]).

The concrete construction of the λ-algebra $\mathcal{P}\omega_{re}$ is relatively familiar and straightforward (see e.g. [Sc76, Ba84]). As is shown in [Sc76], it yields as its Karoubi envelope the category $\mathbf{CLat}_{\mathit{eff}}$ of *effective continuous lattices* and computable maps.

One can now give an interpretation of PCF +choose. in this category. The interpretation of types is given as above but using non-standard interpretations for the ground types:

$$X_{nat} = \mathcal{P}\omega, \qquad X_{bool} = \mathcal{P}2, \qquad X_{unit} = \mathcal{P}1.$$

The interpretation of the constants is reasonably evident: for example, $[\![n]\!] : 1 \to \mathcal{P}\omega$ corresponds to the singleton $\{n\} \in \mathcal{P}\omega$; $[\![\mathsf{choose}]\!]$ corresponds to the set $\omega \in \mathcal{P}\omega$; and $[\![\mathsf{succ}]\!]$ corresponds to the continuous mapping $A \mapsto \{n+1 \mid n \in A\} : \mathcal{P}\omega \to \mathcal{P}\omega$. The interpretation of variables, abstraction and application is then completely standard. Finally, the relation $\sqsubseteq$ for ground types corresponds to set-theoretic inclusion for singleton subsets: if $[\![V]\!]$ corresponds to $\{v\} \in \mathcal{P}\omega$ and $[\![M]\!]$ corresponds to $A \in \mathcal{P}\omega$, then $[\![V]\!] \sqsubseteq [\![M]\!]$ iff $v \in A$. The proof of adequacy for this interpretation is straightforward.

We now claim that *nat* is a universal type for the above interpretation. Clearly all the ground types are definable retracts of *nat*. To define a retraction $[\![nat \to nat]\!] \lhd [\![nat]\!]$, we have to code up the familiar retraction $[\mathcal{P}\omega \to \mathcal{P}\omega] \lhd \mathcal{P}\omega$ in PCF + choose. We give here only a bare outline of how to do this, making free use of pseudocode. Suppose $\langle -, - \rangle$ is an effective pairing operation on N, and $[-]$ is an effective encoding of finite subsets of N as elements of N. Let check be the term $\lambda bn.\mathsf{cond}\ b\ n$ (diverge). Now consider the terms

$$x : nat \ \vdash \lambda y.\ \text{let } \langle [A], r \rangle = x \text{ in } \mathsf{check}\ (\text{for each } a \in A,\ y = a)\ r$$

$$f : nat \to nat \ \vdash \text{let } \langle [A], r \rangle = \mathsf{choose} \text{ in } \mathsf{check}\ (f(\text{let } n = \mathsf{choose} \text{ in } \mathsf{check}\,(n \in A)\, n) = r)\ (\langle [A], r \rangle)$$

For the first term, the intuition is that we regard x as representing a term whose evaluation will give us some non-deterministic choice of an element of the graph of some function $f : \mathcal{P}\omega \to \mathcal{P}\omega$, and our task is to code the (non-deterministic) operation of type $nat \to nat$ whose behaviour on (non-deterministic) arguments y is precisely as given by f. For the second term, given just such an operation, our task is to construct a term of type *nat* that will non-deterministically choose an element of the graph of this operation.

It can be shown that the above terms do define a retraction as required. Since full abstraction and completeness are trivial at type *nat*, we have the following pleasing result, which as far as we are aware has not appeared in the literature before:

Theorem 6.1. *The above interpretation of* PCF_N + choose *in* $\mathbf{CLat}_{\mathit{eff}}$ *is fully abstract and complete.*

6.4. The sequentially realizable functionals

Next, we briefly mention a more complex example in which the existence of a universal type turns out to be extremely useful. We refer the reader to [Lo02] for all the details.

The language PCF^{++} above shows how we can extend PCF with "parallel" operators whilst retaining the purely functional character of the language. ("Functional" here means that the meaning of a closed term of arrow type can be taken to be simply a function between the appropriate sets. More precisely, it means that the term model for the language is a *well-pointed* category: that is, $f = g : X \to Y$ iff $f \circ x = g \circ x$ for all $x : 1 \to X$.) Somewhat surprisingly, it is also possible to extend PCF by adding "sequentially computable" operators not already definable in PCF, and still remain functional in this sense. In fact, it turns out that there is a mathematically natural class of sequentially computable functionals with many attractive properties. These are known in [Lo02] as the *(effective) sequentially realizable* (or SR) functionals, and are closely akin to the *strongly stable* functionals of Bucciarelli and Ehrhard [BE91].

One of the main results of [Lo02] is that in the category constituted by the SR functionals, the object $[N_\perp^N \to N_\perp]$ is universal. Since this object is the denotation of the call-by-value type $(nat \to nat) \to nat$ (call this type $\overline{2}$), we may state this fact as follows: for any call-by-value type σ, its denotation $[\![\sigma]\!]$ in the category of SR functionals is a retract of $[\![\overline{2}]\!]$. This follows easily once we have shown that $[\![\overline{3}]\!] \lhd [\![\overline{2}]\!]$ (where $\overline{3} \equiv \overline{2} \to nat$), which we show by giving a semantic construction of the required retraction. In fact, one half of this retraction is already definable in PCF_V; the other half (the morphism $[\![\overline{2}]\!] \to [\![\overline{3}]\!]$) we may call H.

We may now consider the language obtained by extending PCF_V with a constant $\mathsf{H} : \overline{2} \to \overline{3}$, and stipulating that $[\![\mathsf{H}]\!] = H$. (One can give an operational semantics by means of a compilation to an abstract machine, but we will not give details here.) We now have an instance of the general situation described in this paper: the retraction $[\![(unit \to \overline{2}) \to \overline{2}]\!] \lhd [\![\overline{2}]\!]$ arising from $[\![\overline{3}]\!] \lhd [\![\overline{2}]\!]$ is easily seen to be definable in $PCF_V + \mathsf{H}$, and the model is fully abstract and complete at type $\overline{2}$ even for PCF_V. Thus, the category of SR functionals coincides with the term model for $PCF_V + \mathsf{H}$, and this gives us another mathematical handle on this class of functionals.

In this case, the approach via universal types offers the only known reasonable route to the full abstraction and completeness results. The universality of type $\overline{2}$ is also used in [Lo02] to prove other results: in particular, that various semantic characterizations of the SR functionals do indeed coincide. The idea is that it suffices to show that the monoid of endomorphisms of the object $\overline{2}$ looks the same in various categories; from this it follows for general reasons that the categories coincide, at least as far as the PCF-denotable objects are concerned.

6.5. PCF with control

All the examples considered so far have involved functional languages in the sense that the corresponding category is well-pointed (albeit with a non-standard interpretation of the natural numbers in the case of PCF + choose). For these languages, this boils down to the fact that the *equational context lemma* holds: that is, two closed terms $M, M' : \sigma_1 \to \cdots \to \sigma_r \to \gamma$ are observationally equivalent iff for all closed terms $P_i : \sigma_i$ and values $V : \gamma$ we have

$$MP_1 \ldots P_r \Rightarrow V \;\text{ iff }\; M'P_1 \ldots P_r \Rightarrow V.$$

In this subsection and the next, we consider two examples of languages that are not functional in this sense.

The first example is essentially the language SPCF of Cartwright and Felleisen [CF92]; it is equivalent (in the sense of having an equivalent term model) to the language μPCF introduced in [OS97]. It will be convenient to consider a call-by-value version of this language. Let us add to PCF_V a constant

$$\mathsf{catch} : ((nat \to nat) \to nat) \to nat$$

whose operational behaviour may be described informally as follows. To compute $\mathsf{catch}\ F$, we apply F to a dummy argument $g : nat \to nat$ and observe whether F ever interrogates g with an argument of type nat. If F first calls g with argument n, we "catch" this value and jump out of the computation of $F(g)$, returning $2n$ as the final result of $\mathsf{catch}\ F$. If the computation of $F(g)$ completes successfully without any calls to g, returning the result m, we return $2m+1$ as the result of $\mathsf{catch}\ F$. It is easy to see how the catch operator may be implemented in a programming language with a control mechanism such as exceptions or continuations; indeed, it turns out that $\mathrm{PCF}_V + \mathsf{catch}$ serves well as a prototypical language for "functional programming plus control".

We may formalize the operational semantics of this language in a manner inspired by an implementation of catch using exceptions. For each type σ and each $i \in \mathrm{N}$, let us add an auxiliary constant $\mathsf{e}^i_\sigma : nat \to \sigma$; the idea is that $\mathsf{e}^i_\sigma n$ plays the role of an exception carrying the value n. (We can imagine, if we like, that the same exception $\mathsf{e}^i n$ can manifest itself as a term of any type σ as required; we write $\mathsf{e}^i_\sigma n$ for $\mathsf{e}^i n$ construed as a term of type σ. The role of the superscript i is simply to ensure that we have an infinite supply of distinguishable exceptions.) We declare the constants e^i_σ to be values, so that they can be passed as parameters to functions. Terms of the form $\mathsf{e}^i_\sigma n$ are not themselves declared to be values, but we modify our definition of the evaluation relation $M \Rightarrow U$

to allow U to be either a value V or a term $\mathsf{e}^i_\sigma n$. We may now augment the rules for PCF_V with the following rules for catch:

$$\frac{F\,\mathsf{e}^i_{nat} \Rightarrow \mathsf{e}^i_{nat} n}{\mathsf{catch}\ F \Rightarrow 2n}\ (\mathsf{e}^i \text{ not in } F) \qquad \frac{F\,\mathsf{e}^i_{nat} \Rightarrow m}{\mathsf{catch}\ F \Rightarrow 2m+1}$$

plus a host of other rules to handle the propagation of exceptions, for example:

$$\frac{M \Rightarrow \mathsf{e}^i_{nat} n}{\mathsf{iszero}\ M \Rightarrow \mathsf{e}^i_{bool} n} \qquad \frac{M \Rightarrow \lambda x.M' \quad N \Rightarrow \mathsf{e}^i_\sigma n}{MN \Rightarrow \mathsf{e}^i_\tau n} \qquad \frac{M \Rightarrow \mathsf{e}^i_{\sigma\to\tau} n}{MN \Rightarrow \mathsf{e}^i_\tau n}$$

(With these examples in mind, the reader will easily be able to supply the remaining propagation rules.) In the first rule for catch, the side-condition that e^i does not appear in the term F ensures that the occurrence of e^i_{nat} on the right hand side of $F\,\mathsf{e}^i_{nat} \Rightarrow \mathsf{e}^i_{nat} n$ "arises from" the given occurrence on the left hand side in an obvious sense.

We now declare that we will only really be interested in terms of $\mathrm{PCF}_V + \mathsf{catch}$ — the exceptions are merely auxiliary machinery used to define the operational semantics of this language. Note that no term of $\mathrm{PCF}_V + \mathsf{catch}$ can ever evaluate to an exception.

To get a feel for this language, the reader might like to consider the terms

$$F_0 \equiv \lambda f.\, f\,0 + f\,1, \qquad F_1 \equiv \lambda f.\, f\,1 + f\,0$$

(where we assume + has already been defined), and to verify that catch $F_0 \Rightarrow 0$ but catch $F_1 \Rightarrow 2$. Thus, F_0 and F_1 are observably different, even though they define the same function $[\mathrm{N}_\perp \to \mathrm{N}_\perp] \to \mathrm{N}_\perp$. It follows that in any adequate model for PCF+catch, the denotational elements will need to be something more fine-grained than functions (at least if we wish to work with a standard interpretation of *nat*). It was shown by Cartwright, Curien and Felleisen [CCF94] that Berry and Curien's category of *concrete data structures* (CDSs) and *sequential algorithms* provides an adequate model for PCF+catch. Indeed, it is shown in [CCF94] that the sequential algorithms model of PCF+catch is fully abstract, and in [KCF93] that the *effective* sequential algorithms model is also complete. (These authors work with the call-by-name version, but the results hold equally for the call-by-value version. We will not give a precise definition of sequential algorithms here, though the astute reader will be able to glean more or less what they must be from the facts mentioned below.)

The proofs of the above results in [CCF94, KCF93] work on traditional lines and proceed by showing definability of the finite elements of the model, by induction on types. This technique is rather onerous to apply

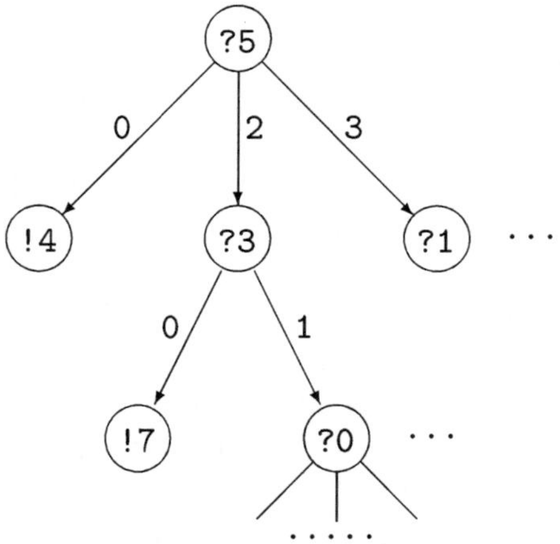


```
fn g =>
  case g(5) of
    0 => output 4
  | 1 => diverge
  | 2 => (case g(3) of
         0 => output 7
       | 1 => (case g(0) of ..)
         ...)
  | 3 => (case g(1) of
         _ => diverge)
   ...
```

Figure 2.1. Part of a decision tree for a sequential strategy

in the case of PCF+catch, and the proof of full abstraction in [CCF94] runs to around 50 pages. It turns out, however, that these results can be proved much more cheaply by exploiting the fact that the type $nat \to nat$ is universal for PCF_V + catch.

Specifically, one first shows that the CDS $[\![(nat \to nat) \to nat]\!]$ is a retract of $[\![nat \to nat]\!]$ in the (effective) sequential algorithms model. The key observation is that a sequential algorithm of type $(nat \to nat) \to nat$ can be represented by a *decision tree*, which can itself be encoded by a function of type $nat \to nat$. The idea behind decision trees should be evident from the example shown in Figure 1. (In fact, sequential algorithms correspond precisely to *irredundant* decision trees — those in which the same question is never asked twice along any path through the tree.) For the encoding in $nat \to nat$, we note that a tree is essentially a partial function from *nodes* in the tree to *labels* on these nodes, which may be questions or answers. A node may be specified by a finite list of natural numbers giving the path to this node; a label is specified by a natural number together with a question/answer tag. It is easy to see that both nodes and labels can be coded up by natural numbers, and hence that a decision tree can be represented by a function $\mathrm{N} \to \mathrm{N}_\perp$.

Likewise, a sequential algorithm of type $(nat \to nat) \to (nat \to nat)$ can be given by an N-indexed forest of such decision trees. By a minor tweak to the above encoding, such forests can also be represented by functions $\mathrm{N} \to \mathrm{N}_\perp$.

Secondly, we show that both halves of this retraction are definable by terms of PCF_V + catch; this is just a little exercise in programming in this language. One half of the retraction is simply the algorithm that plays a decision tree (represented by a function $f : nat \to nat$) against a function $g : nat \to nat$ — this can be done in pure PCF. The other half is an algorithm that extracts the decision tree for a given algorithm $F : (nat \to nat) \to nat$; this makes essential use of catch. As in the example of Section 6.4, it then follows easily that all PCF types are definable retracts of $nat \to nat$.

Thirdly, it is a triviality that at type $nat \to nat$ the sequential algorithms model is fully abstract, and the effective sequential algorithms model is also complete. We are therefore in the situation of Section 4, and so the full abstraction and completeness results follow immediately. (More details of this proof can be found in [Lo02].)

Thus, in the case of PCF_V + catch, the existence of a universal type can be used to great advantage. Indeed, we are inclined to suggest that in this case it even offers a simpler route to the construction of the model in question. The usual definition of CDSs and sequential algorithms is rather cumbersome and requires a fair amount of setting up. Although this approach has its advantages — it is of independent interest to have a concrete description of the objects concerned — we would like to commend the following as an interesting alternative route. Let $\mathcal{B}$ be the λ-algebra corresponding to the object $[\![nat \to nat]\!]$ above, and let $\mathcal{B}_{\mathit{eff}}$ be its effective analogue. It is straightforward enough to construct these λ-algebras concretely from first principles: the underlying set is just the set of partial (recursive) functions $\mathrm{N} \rightharpoonup \mathrm{N}$, and λ-terms are interpreted via encodings of decision trees for the corresponding algorithms as indicated above. (This was first done in [vO99]; for further details see [Lo02].) By taking the Karoubi envelopes of these λ-algebras, we now obtain categories that contain at least the (effectively) sequential CDSs as a full subcategory. This construction gives us all we need in order to define the interpretation of PCF_V+catch and prove adequacy; and as we have seen, full abstraction and completeness are then almost immediate.

6.6. PCF with control and state

In the previous example we saw how universal objects offered a simpler approach to some already known results. We now turn to an example in which our approach leads to some genuinely new (and quite interesting) results. We will concentrate here very much on giving the relevant intuitions — a more formal account of this material will appear in a subsequent paper [Lo03].

We start by giving some background motivation. In the world of sequential algorithms that we have just considered, one takes account of the order of function calls (so that $[\![f\,0 + f\,1]\!] \neq [\![f\,1 + f\,0]\!]$) but not of the number of repetitions (so that $[\![f\,0 + f\,0]\!] = [\![f\,0 * 2]\!]$). This is because sequential algorithms correspond to *irredundant* decision trees — we can skip all repetitions of questions since in this setting we know they will always receive the same answer as before. This corresponds to what one can observe in a language like PCF+catch, with control features (allowing us to distinguish different evaluation orders) but no state (which would be needed to detect a repeated function call). We would now like to tell an analogous story, but in a more fine-grained setting where we take account of both the order and number of function calls, in order to provide a suitable setting for modelling a language with both control features and state. (This way of thinking about control and state as separate "ingredients" of a programming language owes much to Abramsky's work on game semantics, see e.g. [AM99].)

The idea, then, is to give a more refined version of the λ-algebra $\mathcal{B}$, embodying a more intensional notion of computation strategy. The Karoubi envelope of this algebra will then give something more fine-grained than sequential algorithms — let us call the resulting strategies *sequential processes.* We will then be able to give a programming language to match this model, and obtain full abstraction and completeness results.

As a preliminary intuition, let us consider the following ridiculously simple game G. Two participants, Opponent and Player, take turns to name a natural number, with Opponent starting. For example:

Opponent:	5		2		5		...
Player:		3		4		1	

At any stage, either participant may "give up", in which case the game can proceed no further. The play may continue indefinitely. There is no notion of winning or losing.

It might be felt that this game is so boring that nothing of interest can conceivably be said about it. However, what is important about this game is that many much more interesting games can be coded up in it. For example:

1 Chess (Chinese or European). It is easy enough to code up possible moves in chess as natural numbers, and having fixed on a coding convention, you and I could perfectly well play chess just by swapping natural numbers.

2 Any reasonable kind of interaction between two computational agents or coroutines, such as a function and its argument. Indeed, any situation at all in which two participants take turns to

perform finitely describable actions can be regarded as a play of the game G.

We are therefore thinking of G as a kind of "generic" or "universal" game. The idea, then, is that by understanding this game really well, we will be able to get a handle on a very large and rich category of games.

A λ-algebra of strategies. We will now show how to build a λ-algebra inspired by the game G. The first step is to understand the nature of *strategies* for G. A strategy for Opponent may be given in an obvious way as a decision tree in which the nodes are labelled simply by natural numbers (with no distinction between questions and answers): the node labels correspond to Opponent moves, the edge labels to Player moves. A strategy for Player is similar (interchanging the roles of Player and Opponent), except that there is no root node, since the game starts with an Opponent move. (Alternatively, we may think of Player strategies as N-indexed forests of decision trees.)

Whereas in the construction of $\mathcal{B}$ we coded up decision trees as functions $\mathrm{N} \to \mathrm{N}_\perp$, here we will take the trees themselves to be the underlying objects. Let $\mathcal{S}$ be the set of Player strategies as described above; more formally, we may define $\mathcal{S}$ coinductively as the final coalgebra for the domain equation

$$\mathcal{S} \;\cong\; \mathrm{N} \to (\mathrm{N} \times \mathcal{S})_\perp$$

in the category of sets. Given $S \in \mathcal{S}$ and α a finite list of natural numbers, we write $S\!\mid_\alpha \in \mathcal{S}$ for the subforest of S below node α.

We will take $\mathcal{S}$ to be the underlying set of our λ-algebra: the idea is that Player strategies for G can also serve as strategies for all other games of interest. Note in particular that there is no irredundancy condition on strategies; indeed, if Opponent repeats a move, then Player may well decide to respond differently the second time (as in the little dialogue above).

In a programming language with suitable recursive datatypes, clearly we can represent strategies using a type σ of lazy trees. In Standard ML, for example, we may define:

```
datatype strategy = strategy of nat -> (nat * strategy)
```

where `nat` is a type for natural numbers. We may now ask: what does a strategy for an operation (or sequential process) of type $\sigma \to \sigma$ look like (say, in a language like ML)? We will first answer this informally by outlining the way in which a dialogue with such a strategy can proceed. Suppose $P : \sigma \to \sigma$ is some operation, $S : \sigma$ is some strategy, and we wish to obtain the resulting strategy $T = P(S)$. We will let Player play the role of P, and let Opponent play the role of T and also that of the opponent to T. The dialogue may then proceed as follows:

- Opponent starts by playing a move q_0 in the game G against T. That is, Opponent asks about the subtree on branch q_0 in the strategy T.
- Player *either* responds immediately with a move p_0 in G (that is, declares that the label on the node $[q_0]$ in T is p_0), *or* asks about the subtree on some branch m_0 in S.
- Assuming the latter, Opponent must reply by giving the label n_0 on the node $[m_0]$ in S.
- Player now has three options:
  - He may respond with a move p_0 in G.
  - He may ask about the subtree of S on some branch m_1 (possibly $m_1 = m_0$, in which case he is asking the "same" question as before).
  - He may ask about the subtree of $S\mid_{[m_0]}$ on some branch m_1. Here the subforest $S\mid_{[m_0]}$ was "opened up" by the previous move and is now available for exploration by Player.

and so on.

In the case of the last possibility, the subforest $S\mid_{[m_0,m_1]}$ will also be opened up and made available for exploration by subsequent moves. Indeed, at any stage in the game, there will be some prefix-closed set of nodes α of S that we have visited so far, and a corresponding collection of subforests $S\mid_\alpha$ that are currently visible. (Also the forest $S\mid_\epsilon = S$ is already visible at the start of the game, and remains so.) At any stage in the game, Player must either play a move in G or else ask about some branch in one of the visible subforests. The idea should be clear if one bears in mind the possible ways in which an ML program of type `strategy -> strategy` can behave.

Actually, the above description is in need of a small correction. Suppose Player's first move is to ask about branch m_0 of S, and his second move is again to ask about branch m_0 of S. We then have, strictly speaking, two *copies* of $S\mid_{[m_0]}$ opened up, and for subsequent questions about this subforest, we can in principle ask which copy is being explored. Of course, if S is a "pure forest" (*i.e.,* an element of $\mathcal{S}$), it will make no difference, since we will always get the same answer from either copy. However, if the role of S is played by a process with some kind of internal state, we may well find we get different answers from different copies. So strictly speaking, we have at each stage a set of copies of subforests, or *subforest instances*, available for exploration. Note that each time Player asks a question about one of the visible copies, he opens up a new subforest instance (whether or not it has asked the same question before), and so adds exactly one to the stock of visible subforest instances. We may therefore identify these copies by *timestamps*: 0 for the given instance of $\mathcal{S}$ itself, and $1, 2, 3, \ldots$ for the new subforest instances in the order in which they become visible.

Typically, Player will continue exploring subforest instances of S until he has enough information to be able to respond to Opponent's move q_0 in G with a move p_0, signifying that the node $[q_0]$ in T carries the label p_0. Next, Opponent will play a move q_1, asking about branch q_1 of the subforest $T|_{[q_0]}$. Again, Player may continue his exploration of subforest instances of S (all previously visited instances remaining visible), until he is able to respond with a move p_1 in G, and so on. Of course, at any stage in the game, either participant may fail to supply a move (this corresponds to entering a non-terminating computation), in which case the game cannot proceed.

It is now clear how such a strategy can itself be represented by a decision tree. As before, we take labels on nodes to represent Player moves, and labels on edges to represent Opponent moves. Again, there is no root node, since Opponent starts. Each Player move is either an *answer* (*i.e.,* a natural number played in G), or a *question* posed to some subforest instance of S (such a question may be specified by a timestamp identifying the subforest instance interrogated, together with a number giving the branch of this subforest that we are asking for). Thus, node labels (Player moves) may be taken to be elements of $\mathrm{N} + (\mathrm{N} \times \mathrm{N})$. Opponent moves, on the other hand, are either moves in G (natural numbers) or answers to Player questions about node labels in S (again natural numbers). Which of these kinds any Opponent move is may be determined from the context: if the move follows a Player question, it must be a node label in S, otherwise it must be a move in G. This means that no coding is necessary for Opponent moves, and we may take edge labels to be just elements of N.

Using some bijective coding $\mathrm{N} \cong \mathrm{N} + (\mathrm{N} \times \mathrm{N})$, such a decision tree can easily be represented by an element of $\mathcal{S}$ itself. Thus, strategies of type $\mathcal{S} \to \mathcal{S}$ may themselves be represented by elements of $\mathcal{S}$. In the other direction, given strategies R and S we may construe R as representing a strategy of type $\mathcal{S} \to \mathcal{S}$ and so "apply" R to S. (The only subtlety here is that we may at some point encounter an out-of-range timestamp, in which case we should diverge). Armed with the above ideas, it is easy to show along standard lines that $\mathcal{S}$ is indeed a λ-algebra (the proof is very similar to that for $\mathcal{B}$). We also of course have the subalgebra $\mathcal{S}_{\mathit{eff}}$ consisting of the recursive strategies. (A more formal definition of $\mathcal{S}$ and its λ-algebra structure will be given in [Lo03].)

Categories of sequential processes. The Karoubi envelopes $\mathcal{K}(\mathcal{S})$ and $\mathcal{K}(\mathcal{S}_{\mathit{eff}})$ now give us good cartesian closed categories; we may call the morphisms of these categories *(effective) sequential processes.* It is easy to see that these categories provide adequate models for (both versions of) PCF; in fact, we shall consider them as models for PCF_N extended with an additional ground type corresponding to $\mathcal{S}$ itself. Let PCF^*_N be the language obtained by adding to the definition of PCF_N a new ground type ζ representing $\mathcal{S}$ (*not* designated as a program type), together with suitable constants and evaluation rules (the precise details are rendered slightly complicated by the absence of product types in our framework). Then we have an adequate interpretation $[\![-]\!]$ of PCF^*_N in $\mathcal{K}(\mathcal{S})$ such that $[\![\zeta]\!] = \mathcal{S}$ (more accurately, $[\![\zeta]\!] = id \in \mathcal{S}$); similarly for $\mathcal{S}_{\mathit{eff}}$. It is easy to see that both these interpretations are fully abstract at type ζ, and that the interpretation in $\mathcal{K}(\mathcal{S}_{\mathit{eff}})$ is complete at type ζ; also that *nat*, *bool*, *unit* are PCF^*_N-definable retracts of ζ. The λ-algebra structure of $\mathcal{S}, \mathcal{S}_{\mathit{eff}}$ gives us a retraction $[\![\zeta \to \zeta]\!] \lhd [\![\zeta]\!]$, of which one half (the morphism *Apply*) is already definable in PCF^*_N.

Now let us add to PCF^*_N a constant

$$\mathsf{lambda} : (\zeta \to \zeta) \to \zeta$$

and stipulate that $[\![\mathsf{lambda}]\!]$ is the other half of the above retraction. It is then immediate that $\mathcal{K}(\mathcal{S}_{\mathit{eff}})$ is a complete model for $\mathrm{PCF}^*_N + \mathsf{lambda}$, and that both models are fully abstract in the sense that $[\![M]\!] = [\![N]\!]$ iff $[\![C[M]]\!] = [\![C[N]]\!]$ for all contexts $C[-]$ of program type. That is, the models would be fully abstract if we could give an operational semantics for $\mathrm{PCF}^*_N + \mathsf{lambda}$ with respect to which our interpretations are adequate.

This naturally raises the question: is there any reasonable operational semantics for $\mathrm{PCF}^*_N + \mathsf{lambda}$? As we have seen, lambda is an operation which extracts the computation strategy behind a given program of type $\zeta \to \zeta$ — but is this operation "computable" in any natural sense? Fortunately, the answer is yes: in fact, lambda is programmable in a language with continuations and local state (such as New Jersey ML), so one can at least give an operational semantics for $\mathrm{PCF}^*_N + \mathsf{lambda}$ indirectly via a translation into such a language. We can then think of lambda as a primitive, and show that the above denotational semantics for $\mathrm{PCF}^*_N + \mathsf{lambda}$ is indeed adequate.

We have, in fact, implemented lambda in New Jersey ML as part of a program called STRATAGEM [STR], which may be downloaded from the author's web page. This program allows the user to extract the computation strategy behind a program of any type, and to interactively explore and manipulate this strategy in a variety of interesting ways. The whole system is based on the universality of the type ζ, and serves as a showcase for many of the above ideas.

We therefore have, in principle, a programming language whose fully abstract and complete model is essentially $\mathcal{K}(\mathcal{S}_{\mathit{eff}})$. Still, lambda is hardly a natural language primitive from a programming point of view: it would be nice to have a more programmer-friendly characterization of a language that was equivalent in expressive power to $\mathrm{PCF}_N^* + \mathsf{lambda}$. In fact, it turns out that the language $\mathcal{L}$ = "PCF + continuations + local state", if appropriately defined, can both implement lambda and be adequately translated into $\mathrm{PCF}_N^* + \mathsf{lambda}$, so that the term models for these languages are equivalent (as far as the types of PCF_N^* are concerned). We therefore have that $\mathcal{K}(\mathcal{S}_{\mathit{eff}})$ is essentially the fully abstract, complete model of $\mathcal{L}$. The precise definition of $\mathcal{L}$, and the details of the syntactic translations between $\mathcal{L}$ and $\mathrm{PCF}_N^* + \mathsf{lambda}$, will be presented in [Lo03].

To summarize, starting from an intuitive concept of computation strategy we have constructed a λ-algebra $\mathcal{S}_{\mathit{eff}}$, and hence a category $\mathcal{K}(\mathcal{S}_{\mathit{eff}})$, and then found a programming language $\mathcal{L}$ to match it, guided by the existence of a universal type. Since $\mathcal{S}_{\mathit{eff}}$ is mathematically a very tractable object, we therefore have a very good grasp of the term model for $\mathcal{L}$. In our view, this way of using models to inspire our choice of languages is a very significant part of the usefulness of denotational semantics: it helps us to identify languages for which observational equivalence and definability are simple to characterize, and which therefore lend themselves to the design of clean and principled program logics.

We emphasize that in this situation we are able to reap all the benefits of having a denotational semantics without ever giving a concrete description of the objects of $\mathcal{K}(\mathcal{S}_{\mathit{eff}})$. In fact, it turns out that this category is essentially the same as (*i.e.,* has a large full subcategory in common with) one of the categories of games described in [AM99], though we only discovered this after obtaining the above results. Specifically, our category agrees with the intensional category of games and effective strategies without well-bracketing or innocence constraints; we therefore have that $\mathcal{L}$ is a complete language for this category of games. Although the idea that this category corresponds to "PCF + control + state" is explicit in [AM99], no precise language matching this model was given there, and so this appears to be a new result.

Finally, we remark on a curious feature of this example: there is a universal type for $\mathrm{PCF}_N^* + \mathsf{lambda}$, but there is no universal *simple* type (that is, a type built up from *unit*, *bool*, *nat* using $\to$). So the addition of a type ζ to our language is quite essential for our purposes. This is perhaps surprising, because in the setting of weaker languages such as PCF (or even PCF+catch), we are accustomed to the idea that recursive datatypes such as ζ can be expressed as retracts of simple types. The explanation, of course, is that a pair of terms (F, G) defining a retraction $\zeta \lhd \sigma$ in PCF + catch need not yield a retraction in PCF + lambda: the observational equivalence in the latter language may be sufficiently fine-grained that we no longer have $G \circ F \approx id$.

7. Conclusions and further directions

We end by drawing together some conclusions from the discussion in this paper, and pointing out a few further ramifications of the ideas.

7.1. Full abstraction and completeness results

In this paper we have discussed an approach to denotational semantics which exploits the existence of universal types in many computational settings, and shown how this leads to full abstraction and completeness results for various programming languages. The whole approach is somewhat opportunistic, in that it takes advantage of universal types when these happen to exist; it is quite inapplicable in situations where there is no universal type (e.g. the language PCF). It could be objected that the existence of a universal type is a rather *ad hoc* property of a programming language; however, as our examples show, a remarkably large number of interesting programming languages do possess universal types, so our approach works well in a large number of cases. As we have seen, an approach via universal types is sometimes simpler than other approaches (as with PCF+catch), and sometimes harder (as with PCF^{++}) — though even in the latter cases it offers an interesting alternative route to the results. Moreover, we have seen that our approach can inspire interesting new results (as with PCF+choose, PCF + H and $\mathrm{PCF}_N^* + \mathsf{lambda}$).

Besides giving a method for proving full abstraction and completeness results, we have seen that universal types often give an attractive way of constructing the categories of interest in the first place. Our most dramatic example was the category $\mathcal{S}$, for which this was the *only* construction we gave — we were able to obtain all the results we wanted without ever giving a concrete description of the category. One might feel that this is cheating in some way — certainly, for a well-rounded understanding of the category, an explicit description of the objects and morphisms seems desirable — but it is nonetheless interesting to note that technically we can get away without it, and that it may be simpler to do so.

It is also perhaps of interest that our approach achieves some uniformity in its treatment of a diverse collection of instances: in each case we simply have to check the five conditions listed at the end of Section 4. From an expository point of view, our approach thus offers an efficient way of covering a wide range of scenarios in denotational semantics.

We have concentrated in this paper on cartesian closed categories and their connection with λ-algebras. However, as mentioned in Section 2, there are other correlations of this kind that one might consider. For instance, λ-algebras with a certain *least fixed point* property give rise to categories with initial algebras for "representable endofunctors", and hence to models for inductive datatypes (this is a straightforward abstraction of the ideas in [Sc76]). All the examples of λ-algebras A that we have considered carry a well-behaved partial ordering, and moreover have the property that every representable function $A \to A$ has a least fixed point, so it would be straightforward to extend all our results to versions of our languages with inductive types. Another example is the correlation between *linear* λ-algebras and symmetric monoidal closed categories. It is easy enough to give a "linear" version of many of the ideas in this paper, and this is likely to play a useful role in the study of programming languages with a linear flavour (e.g. languages involving objects with state).

7.2. A notion of expressive equivalence

On a more conceptual level, the point of view we have advocated suggests one possible notion of when two programming languages (or models) are essentially *equivalent* in terms of their expressive power. The idea behind our whole approach is that two categories match up well iff they have the same Karoubi envelope. (The Karoubi envelope $\mathcal{K}(\mathcal{C})$ of a general category $\mathcal{C}$ has as objects all morphisms $a : X \to X$ in $\mathcal{C}$ such that $a^2 = a$, and as morphisms $a \to b$ all morphisms f in $\mathcal{C}$ such that $b \circ f \circ a = f$.) In this paper we have emphasized the case where one category is a term model for a programming language, and the other is the monoid given by a λ-algebra, but one can propose the same condition more generally as a notion of "expressive equivalence".

Let us say that a general category $\mathcal{C}$ is a *λ-algebra* iff $\mathcal{K}(\mathcal{C})$ is cartesian closed; it is then natural to say that $\mathcal{C}$ and $\mathcal{D}$ are *equivalent* as λ-algebras iff $\mathcal{K}(\mathcal{C})$ and $\mathcal{K}(\mathcal{D})$ are equivalent as categories. If both categories are syntactically constructed from programming languages, this gives us a reasonable notion of expressive equivalence. (For instance, the call-by-name and call-by-value variants of a programming language will typically be equivalent in this sense: see [Lo95, Chapter 6].) If one of the categories is semantically constructed, the equivalence amounts to a full abstraction and completeness result. As an example, we have that the following λ-algebras are equivalent:

- (The term model for) the language $\mathrm{PCF}_V + \mathsf{catch}$.
- (The term model for) its call-by-name analogue $\mathrm{PCF}_N + \mathsf{catch}$.
- (The monoid arising from) the untyped λ-algebra $\mathcal{B}_{\mathit{eff}}$.
- The category of effectively sequential CDSs and effective sequential algorithms.

(Note that this notion of "equivalence as λ-algebras" is stronger than the notion of "equivalence as partial combinatory algebras" introduced in [Lo99]. For example, our untyped structures $\mathcal{B}$ and $\mathcal{S}$ are equivalent as combinatory algebras, but not as λ-algebras.)

7.3. Program logics

The usefulness of universal types does not end with the proofs of full abstraction and completeness results. Indeed, in the author's view, one of the chief reasons why these results are themselves interesting is that they offer a stepping-stone to the design of clean logics for reasoning about programs, and here again universal types have a valuable part to play.

Let us recall briefly the approach to program logics advocated e.g. in [LP97]. Let $\mathcal{L}$ be a programming language as in Section 3 above, and consider a many-sorted first-order predicate logic $\mathbf{J}_{\mathcal{L}}$ given as follows:

- Sorts of $\mathbf{J}_{\mathcal{L}}$ are just types of $\mathcal{L}$.
- Terms of sort σ in $\mathbf{J}_{\mathcal{L}}$ are just terms of type σ in $\mathcal{L}$.
- For each type σ we have a binary relation symbol $=_\sigma$, and for each program type σ we have a unary relation symbol $\Downarrow_\sigma$.
- Formulae are built up as usual from atomic formulae using the connectives $\wedge, \vee, \Rightarrow, \forall, \exists$.

We would like an interpretation of this logic in terms of purely operational concepts, so that a programmer familiar with $\mathcal{L}$ can understand what sentences of $\mathbf{J}_{\mathcal{L}}$ mean without knowing anything about denotational semantics. One such interpretation (for closed formulae of $\mathbf{J}_{\mathcal{L}}$) is the following:

- The relations $=$ and $\Downarrow$ are interpreted as *observational equivalence* and *termination* respectively.
- The propositional connectives have their usual truth-table interpretation.
- Bound variables are understood as ranging over closed terms of $\mathcal{L}$: thus, a formula $\forall x : \sigma.\phi$ is interpreted as saying that $\phi[M/x]$ holds for all closed terms $M : \sigma$.

With this interpretation, the logic $\mathbf{J}_{\mathcal{L}}$ provides a powerful specification language for expressing correctness properties of programs, much in the spirit of Extended ML [KST97].

We would now like to design a sound and reasonably complete proof system for the logic with this interpretation. Clearly, the standard rules of classical logic will suffice as inference rules, but the problem is to find a good set of axioms which suffice for proving interesting properties of programs. In view of Gödel's Incompleteness Theorem, one cannot hope to provide a complete axiomatization, but one might at least hope for a *relative completeness* result, saying for instance that our axiom system was "as complete as Peano arithmetic".

Denotationally, our interpretation of $\mathbf{J}_{\mathcal{L}}$ obviously coincides with the usual Tarskian interpretation of first order logic over the naive set-theoretic model in which we take $[\![\sigma]\!]$ to be just the set of closed terms $M : \sigma$ modulo observational equivalence. Our goal is to provide a set of axioms that suffice to "pin down" what this structure is up to isomorphism (as well as the interpretation of particular terms within this structure). The hope is that a good fully abstract and complete model for $\mathcal{L}$ should give us a more semantic characterization of this structure, giving us a better mathematical handle on it and hence yielding a good choice of axioms for achieving our goal.

We may now go a stage beyond the ideas in [LP97]. In view of the above caveat about incompleteness, we had better limit our ambitions to pinning down the structure under the assumption that the set $[\![nat]\!]$ (say) is *standard* — that is, canonically isomorphic to the usual set $\mathrm{N}_{\perp}$. Here, the existence of a universal type σ can be very useful: once we have given enough axioms to pin down the structure of $[\![\sigma]\!]$ (under the above assumption), and also to pin down the idempotents on $[\![\sigma]\!]$ corresponding to the retractions $[\![\tau]\!] \lhd [\![\sigma]\!]$, we have said enough to determine the model up to isomorphism, and our axiomatization will be relatively complete. Since the universal type is typically a relatively low type, it is generally not too hard to come up with axioms that suffice to pin down its structure.

The above approach to program logics forms the basis of a current project in which we are developing a sound and relatively complete proof system for some large fragments of ML [Prop]. In this project, we identify three sublanguages of New Jersey ML of increasing expressive power: a language $\mathcal{L}_1$ equivalent to PCF + H; a language $\mathcal{L}_2$ equivalent to PCF + catch; and a language $\mathcal{L}_3$ equivalent to PCF* + lambda. (The notion of equivalence here is the one described in Section 7.2; in fact, one can regard the languages $\mathcal{L}_i$ simply as sugared versions of the corresponding "toy language".) Our purpose is to provide program logics along the above lines for these languages, to encode these in the Isabelle theorem prover [Pa94], and to develop some support for verifying properties of programs in these logics. Although the languages $\mathcal{L}_i$ are much more complex in their detailed description than the corresponding toy languages, the above equivalences mean that their term models are well understood, and so our method scales up successfully. (Indeed, a crucial ingredient in our methodology is the idea that these considerations should be used to help *determine* the choice of fragments of $\mathcal{L}_i$ — we then know in advance that we are dealing with languages for which we have a good semantic understanding.) Moreover, in each of these cases there is a universal type, which enables us to come up with a simple but relatively complete axiom system in the manner outlined above.

Finally, we are hopeful that universal types may be of further assistance when we come to design proof tactics and decision algorithms for restricted fragments of the logics, since we can concentrate our attention on designing tactics and algorithms that work for the universal type. It will be interesting to see whether this is indeed the case.

References

[AM99] Abramsky, S. and McCusker, G., Game semantics, In: H. Schwichtenberg and U. Berger (editors), *Computational Logic: Proceedings of the 1997 Marktoberdorf summer school*, Springer-Verlag, 1999, 1–56.

[Ba84] Barendregt, H. P., *The Lambda Calculus: Its Syntax and Semantics*, North-Holland 1984.

[BE91] Bucciarelli, A. and Ehrhard, T., Sequentiality and strong stability, In: *Proc. 6th Annual Symposium on Logic in Computer Science*, IEEE Computer Soc. Press, 1991, 138–145.

[CCF94] Cartwright, R., Curien, P.-L. and Felleisen, M., Fully abstract semantics for observably sequential languages, *Information and Computation* **111(2)** (1994), 297–401.

[CF92] Cartwright, R. and Felleisen, M., Observable sequentiality and full abstraction, In: *Proc. 19th Symp. Principles of Programming Languages*, ACM Press, 1992, 328–342.

[HO00] Hyland, J. M. E. and Ong, C.-H. L., On full abstraction for PCF: I, II and III, *Information and Computation* **163** (2000), 285–408.

[KCF93] Kanneganti, R., Cartwright, R. and Felleisen, M., SPCF: its model, calculus, and computational power, In: *Proc. REX Workshop on Semantics and Concurrency, Lecture Notes in Computer Science 666*, Springer-Verlag, 1993, 318–347.

[KST97] Kahrs, S. and Sannella, D. and Tarlecki, A., The definition of Extended ML: a gentle introduction, *Theor. Comp. Sci.* **173** (1997).

[K082] Koymans, K., Models of the lambda calculus, *Information and Control* **52** (1982), 306–332.

[Lo95] Longley, J. R., *Realizability Toposes and Language Semantics*, PhD thesis, University of Edinburgh, 1995.

[Lo99] —, Matching typed and untyped realizability, In: Proc. Workshop on Realizability, Trento, published as *Electronic Notes in Theoretical Computer Science* **23(1)**, Elsevier, 1999.

[Lo02] —, The sequentially realizable functionals, *Annals of Pure and Applied Logic* **117(1)** (2002), 1–93.

[Lo03] —, A category of sequential processes, In preparation, 2003.

[LP97] Longley, J. R. and Plotkin, G. D., Logical full abstraction and PCF, In: J. Ginzburg (editor), *Tbilisi Symposium on Language, Logic and Computation*, SiLLI/CSLI, 1997, 333–352.

[OS97] Ong, C.-H. L. and Stewart, C. A., A Curry-Howard foundation for functional computation with control, In: *Proc. 24th Symp. on Principles of Programming Languages*, ACM Press, 1997, 215–227.

[Pa94] Paulson, L. C., *Isabelle: A generic theorem prover*, Vol. 828 of *Lecture Notes in Computer Science*, Springer-Verlag, 1994.

[Pl77] Plotkin, G. D., LCF considered as a programming language, *Theoretical Computer Science* **5** (1977), 223–255.

[Pl78] —, $\mathbb{T}^\omega$ as a universal domain, *Journal of Computer and System Sciences* **17** (1978), 209–236.

[Prop] Fourman, M. P., Fleuriot, J. D. and Longley, J. R., A proof system for correct program development. Case for support for EPSRC grant GR/N64571, available from the author's home page, 2000.

[Sa76] Sazonov, V. Yu., Degrees of parallelism in computations, In: *Mathematical Foundations of Computer Science 1976*, Vol. 45 of *Lecture Notes in Computer Science*, Springer-Verlag, 1976, 517–523.

[Sc76] Scott, D. S., Data types as lattices, *SIAM Journal of Computing* **5** (1976), 522–587.

[Sc80] —, Lambda calculus: some models, some philosophy, In: J. Barwise, H.J. Keisler and K. Kunen (editors), *The Kleene Symposium*, North-Holland, 1980, 223–266.

[Sc93] —, A type-theoretical alternative to ISWIM, CUCH, OWHY, *Theoretical Computer Science* **121** (1993), 411–440. First written in 1969 and widely circulated in unpublished form since then.

[Si90] Sieber, K., Relating full abstraction results for different programming languages, In: *Proc. 10th Conference on Foundations of Software Technology and Theoretical Computer Science, Bangalore*, Vol. 472 of *Lecture Notes in Computer Science*, Springer-Verlag, 1990.

[STR] STRATAGEM, for SML of New Jersey, Source code and user documentation available from the author's home page, 2001.

[vO99] van Oosten, J., A Combinatory Algebra for Sequential Functionals of Finite Type, In: S.B. Cooper and J.K. Truss (editors), *Models and Computability*, Cambridge University Press, 1999, 389–406.

Chapter 3

RELATIONAL REPRESENTATIONS OF HYPER-CONTINUOUS LATTICES*

Xiao-Quan Xu
Department of Mathematics, Sichuan University
Chengdu, Sichuan 610064, China
Department of Mathematics, Jiangxi Normal University
Nanchang, Jiangxi 330027, P.R. China
xiqxu@public.nc.jx.cn

Ying-Ming Liu
Department of Mathematics, Sichuan University
Chengdu, Sichuan 610064, China
ymliu@scu.edu.cn

Abstract The relational representation and the intrinsic characterization of hyper-continuous lattices are given.

Keywords: Relational representation, hyper-continuous lattice, regular relation, finitely regular relation

The regularity of binary relations were first characterized by Zareckiĭ. In [Za63] he proved the following remarkable result: a binary relation ρ on a set X is regular if and only if the complete lattice $(\Phi_\rho(X), \subseteq)$ is completely distributive, where

$$\Phi_\rho(X) = \{\rho(A) : A \subseteq X\}, \ \rho(A) = \{y \in X : \exists a \in A \text{ with } (a, y) \in \rho\}.$$

Further criteria for regularity were given by Markowsky[Ma72] and Schein [Sc76] (see also Yang[Ya69] and Bandelt[Ba80, Ba82]).

*Project supported by the National Natural Science Foundation of China, the SFEM of China and the Natural Science Foundation of Jiangxi Province, China

GQ Zhang et al. (eds.), Domain Theory, Logic and Computation, 65-74.

In the last twenty years, the generalizations of continuous lattices (domains) have attracted a considerable amount of attention. In the words of the authors of [GLS83]: "there is the attempt to carry as much as possible of the theory of continuous lattices to as general an ordered structure as possible." In [GL81] hyper-continuous lattices and generalized continuous lattices were introduced, which are among the most successful such generalizations.

In this paper we give the relational representation and the intrinsic characterization of hyper-continuous lattices. Also in the paper, the intrinsic characterization of regular relations is obtained.

1. Preliminaries

In the paper, **Set** denotes the class of all sets. The class of all complete lattices is denoted by **Com**. For any $L \in$**Com**, $x \in L$ and $A \subseteq L$, let $\uparrow x = \{y \in L : x \leq y\}$ and $\uparrow A = \bigcup_{a \in A} \uparrow a$; dually define $\downarrow x$ and $\downarrow A$. The topology generated by the collection of sets $L \setminus \downarrow x$ (as subbasic open subsets) is called the *upper topology* on L and is denoted by $\upsilon(L)$. For two complete lattices L_1 and L_2, the symbol $L_1 \cong L_2$ means that L_1 is isomorphic to L_2. For any $X \in$ **Set**, let $X^{(<\omega)} = \{F \subseteq X : F$ is finite and nonempty$\}$.

For a set X, we call ρ a binary relation on X, written $\rho : X \rightharpoonup X$, if $\rho \subseteq X \times X$. Let $\mathcal{B}(X)$ denote the set of all binary relations on X and let **Rel**$= \bigcup_{X \in \mathbf{Set}} \mathcal{B}(X)$. For $\rho, \tau \in \mathcal{B}(X)$, define $\tau \circ \rho = \{(x, z) \in X \times X : \exists y \in X$ with $(x, y) \in \rho$ and $(y, z) \in \tau\}$. The relation $\tau \circ \rho : X \rightharpoonup X$ is called the *composition* of ρ and τ.

For a binary relation ρ on a set X, define

$$\begin{aligned}
\rho^{-1} &= \{(y, x) \in X \times X : (x, y) \in \rho\}, \\
\rho' &= X \times X \setminus \rho, \\
\rho^* &= (\rho^{-1} \circ \rho' \circ \rho^{-1})', \\
\rho(A) &= \{y \in X : \exists a \in A, (a, y) \in \rho\} \ \ (\forall A \subseteq X), \text{and} \\
\Phi_\rho(X) &= \{\rho(A) : A \subseteq X\}.
\end{aligned}$$

Clearly, $(\Phi_\rho(X), \subseteq)$ is a complete lattice in which the join operation $\vee$ is the set union operator $\cup$. But the meet operation $\wedge$ in $(\Phi_\rho(X), \subseteq)$ is not the set intersection operator $\cap$ in general.

Definition 1.1. Suppose that p is a property about complete lattices and s a property about binary relations. Let $\mathbf{P} = \{L \in \mathbf{Com} : L$ has the property $p\}$ and $\mathbf{S} = \{\rho \in \mathbf{Rel} : \rho$ has the property $s\}$. We call the relations of s type is a representation of the complete lattices of p type if the following two conditions are satisfied:

(1) $\forall \rho : X \rightharpoonup X, \rho \in \mathbf{S} \Leftrightarrow (\Phi_\rho(X), \subseteq) \in \mathbf{P}$;

(2) $\forall L \in \mathbf{P}, \exists \rho \in \mathbf{S}$ with $L \cong (\Phi_\rho(X), \subseteq)$.

In order to make sense of the s type relational representation of p type lattices, the properties p and s need to be explicitly defined.

For a complete lattice L, define a binary relation $\ll$ (called the *way below relation*) on L by $x \ll y \Leftrightarrow \forall D \in \mathcal{D}(L), y \leq \vee D \Rightarrow x \leq d$ for some $d \in D$. L is called *continuous* if $x = \vee\{u \in L : u \ll x\}$ for all $x \in L$. L is called completely distributive if the following condition holds for all $\{x_{ij} : j \in J_i, i \in I\} \subseteq L$:

$$\bigwedge_{i\in I} \bigvee_{j\in J_i} x_{ij} = \bigvee_{\varphi\in\prod J_i} \bigwedge_{i\in I} x_{i\varphi(i)}.$$

For $L \in \mathbf{Com}$, define a binary relation $\lhd$ (called the *completely below relation*) on L by $x \lhd y \Leftrightarrow \forall A \subseteq L, y \leq \vee A \Rightarrow \exists a \in A$ with $x \leq a$. Let $\Downarrow x = \{u \in L : u \lhd x\}$ and $\Uparrow x = \{v \in L : x \lhd v\}$. By the definition of $\lhd$ we get the following

Lemma 1.1. *Let $L \in$* **Com**, *$x \in L$ and $A \subseteq L$. Then $x \lhd \vee A \Rightarrow \exists a \in A$ with $x \lhd a$. Therefore, $\Downarrow \vee A = \bigcup_{a\in A} \Downarrow a$.*

Lemma 1.2. (Raney[Ra53]) *Let $L \in$* **Com**. *Then*

(1) *L is completely distributive $\Leftrightarrow \forall x \in L, x = \vee\{y \in L : y \lhd x\}$.*

(2) (Interpolation Property) *$\forall x, y \in L, x \lhd y \Rightarrow \exists z \in L$ with $x \lhd z \lhd y$.*

2. Regular representations of completely distributive lattices

Definition 2.1. A binary relation ρ on X is said to satisfy the *interpolation property* if the following condition holds for all $x, y \in X$:
(INT) $x\rho y \Rightarrow \exists z \in X$ with $x\rho z\rho y$.

For $X \in \mathbf{Set}$, it is easy to see that $(\mathcal{B}(X), \circ)$ is a semigroup with the unit $\Delta(X) = \{(x, x) : x \in X\}$. For $\rho \in \mathcal{B}(X)$, if $\rho^2 = \rho \circ \rho = \rho$, then ρ is called *idempotent*.

Definition 2.2. A binary relation ρ on a set X is called *regular* if ρ is a regular element in the semigroup $(\mathrm{B}(X, X), \circ)$, i.e., $\exists \sigma \in \mathcal{B}(X)$ such that $\rho \circ \sigma \circ \rho = \rho$.

Theorem 2.1. (Zareckiĭ[Za63], Markowsky[Ma72], Schein[Sc76]) *For a binary relation ρ on a set X, the following conditions are equivalent:*

(1) *ρ is regular;*

(2) *$(\Phi_\rho(X), \subseteq)$ is a completely distributive lattice;*

(3) $\forall M \in \Phi_\rho(X), M = \bigcup_{y \in M} \bigwedge_{y \in N \in \Phi_\rho(X)} N;$

(4) $\rho = \rho \circ \rho^* \circ \rho;$

(5) $\rho \subseteq \rho \circ \rho^* \circ \rho.$

Now we give an intrinsic characterization of regular relations.

Theorem 2.2. *For a binary relation ρ on a set X, the following two conditions are equivalent:*

(1) *ρ is regular;*

(2) $\forall x, y \in X, (x, y) \in \rho \Rightarrow \exists u, v \in X$ *such that*

(a) $(x, v) \in \rho, (u, y) \in \rho$, *and*

(b) *for all* $s, t \in X, (s, v) \in \rho$ *and* $(u, t) \in \rho \Rightarrow (s, t) \in \rho;$

Proof. (1) $\Rightarrow$ (2): By the regularity of ρ, there is a $\sigma : X \rightharpoonup X$ with $\rho \circ \sigma \circ \rho = \rho$. $\forall x, y \in X$, if $(x, y) \in \rho$, then $(x, y) \in \rho \circ \sigma \circ \rho$. Whence there exist $u, v \in X$ such that $(x, v) \in \rho, (v, u) \in \sigma$ and $(u, y) \in \rho$. Hence we get the condition (a) . Now we check the condition (b). $\forall s, t \in X$, if $(s, v) \in \rho$ and $(u, t) \in \rho$, then by $(v, u) \in \sigma$, we have $(s, t) \in \rho \circ \sigma \circ \rho = \rho$.

(2) $\Rightarrow$ (1): Define a binary relation $\rho^\dagger : X \rightharpoonup X$ by

$$(p, q) \in \rho^\dagger \Leftrightarrow \forall c, d \in X, (c, q) \in \rho \text{ and } (p, d) \in \rho \Rightarrow (c, d) \in \rho.$$

We show that $\rho \circ \rho^\dagger \circ \rho = \rho$. $\forall x, y \in X$, if $(x, y) \in \rho$, then by (2), $\exists u, v \in X$ which satisfy the conditions (a) and (b) in (2). By the definition of $\rho^\dagger$, $(u, v) \in \rho^\dagger$; and hence by $(x, v) \in \rho$ and $(u, y) \in \rho$, $(x, y) \in \rho \circ \rho^\dagger \circ \rho$. On the other hand, if $(x, y) \in \rho \circ \rho^\dagger \circ \rho$, then there exist $p, q \in X$ such that $(x, q) \in \rho, (q, p) \in \rho^\dagger$ and $(p, y) \in \rho$. By the definition of $\rho^\dagger$, $(x, y) \in \rho$. Hence $\rho = \rho \circ \rho^\dagger \circ \rho$. Thus ρ is regular.

Theorem 2.3. *For a complete lattice L, the following conditions are equivalent:*

(1) *L is completely distributive;*

(2) *there is an idempotent relation $\rho : X \rightharpoonup X$ with $L \cong (\Phi_\rho(X), \subseteq)$;*

(3) *there is a regular relation* $\rho : X \rightharpoonup X$ *with* $L \cong (\Phi_\rho(X), \subseteq)$;

(4) *the relation* $\not\leq$ *on* L *is regular*;

(5) $\forall x, y \in L, x \not\leq y \Rightarrow \exists u, v \in L$ *such that*

(a) $x \not\leq v, u \not\leq y$, *and*

(b) $\forall z \in L, u \leq z$ *or* $z \leq v$.

Proof. (1) $\Rightarrow$ (2): Let $X = L$ and $\rho = \rhd : X \rightharpoonup X$, that is, $(x,y) \in \rho \Leftrightarrow y \lhd x$. Then ρ is transitive; and hence by Lemma 1.2(2), ρ is idempotent. For $A \subseteq X$, by Lemma 1.1, $\rho(A) = \bigcup_{a \in A} \Downarrow a = \Downarrow \vee A$. Hence $(\Phi_\rho(X), \subseteq) = (\{\Downarrow x : x \in L\}, \subseteq)$. Define a mapping $\varphi : L \to (\Phi_\rho(X), \subseteq)$ by $\varphi(x) = \Downarrow x$ $(\forall x \in L)$. By Lemma 1.2(1) it is easy to see that φ is a lattice isomorphism. Thus $L \cong (\Phi_\rho(X), \subseteq)$.

(2) $\Rightarrow$ (3): Suppose that $\rho : X \rightharpoonup X$ is idempotent. Then $\rho \circ \rho \circ \rho = \rho \circ \rho = \rho$. Hence ρ is regular.

(3) $\Rightarrow$ (1): By Theorem 2.1.

(1) $\Leftrightarrow$ (4): For $A \subseteq L, \not\leq (A) = \{y \in L : \exists a \in A \text{ with } a \not\leq y\} = L \setminus \uparrow \vee A$. Hence $(\Phi_{\not\leq}(L), \subseteq) = (\{L \setminus \uparrow x : x \in L\}, \subseteq) \cong L$. By Theorem 2.1 we know that the conditions (1) and (4) are equivalent.

(4) $\Rightarrow$ (5): $\forall x, y \in L$ with $x \not\leq y$, by the regularity of $\not\leq$ and Theorem 2.2, there is a $(u, v) \in L \times L$ such that

(a) $x \not\leq v, u \not\leq y$, and

(b′) $\forall s, t \in L, u \not\leq t$ and $s \not\leq v \Rightarrow s \not\leq t$.

For $z \in L$, let $s = t = z$. Then by (b′), we have $u \leq z$ or $z \leq v$. Therefore, u and v satisfy the condition (b) in (5).

(5) $\Rightarrow$ (4): For all $x, y \in L$ with $x \not\leq y$, by (5), $\exists u, v \in L$ such that

(a) $x \not\leq v, u \not\leq y$, and

(b) $\forall z \in L, u \leq z$ or $z \leq v$.

For all $(s,t) \in L \times L$, $s \not\leq v$ and $u \not\leq t \Rightarrow s \not\leq v$ and $t \leq v$ (by (b)) $\Rightarrow s \not\leq t$. Hence $\rho = \not\leq$ satisfies Condition (2) in Theorem 2.2. Therefore, by Theorem 2.2, $\not\leq$ is regular.

Condition (5) in Theorem 2.3 is the intrinsic characterization of complete distributivity. It was first obtained by Raney[Ra60]. Theorem 2.3 shows that the intrinsic characterization of complete distributivity is another expression of the regularity of $\not\leq$ (see Bandelt[Ba80]).

3. Finitely regular representations of hyper-continuous lattices

In this section we give the relational representation and the intrinsic characterization of hyper-continuous lattices.

Definition 3.1. Let L be a complete lattice. Define a binary relation $\prec_{v(L)}$ on L by $x \prec_{v(L)} y \Leftrightarrow y \in \mathrm{int}_{v(L)} \uparrow x$. A complete lattice L is called *hyper-continuous* if for all $x \in L, x = \bigvee\{z \in L : z \prec_{v(L)} x\}$.

Definition 3.2. A binary relation ρ on a set X is called *finitely regular* if for all $(x, y) \in \rho, \exists u \in x$ and $\{v_1, v_2, \cdots, v_k\} \in X^{(<\omega)}$ such that

(i) $(u, y) \in \rho, (x, v_i) \in \rho$ for each $i \in \{1, 2, \cdots, k\}$, and

(ii) for all $\{s_1, s_2, \cdots, s_k\} \in X^{(<\omega)}$ and $t \in X$, $(u, t) \in \rho$ and $\{(s_1, v_1), (s_2, v_2), \ldots, (s_k, v_k)\} \in \rho \Rightarrow (s_j, t) \in \rho$ for some $j \in \{1, 2, \cdots, k\}$.

Remark 3.1. *(1)* Condition *(ii)* in Definition *3.1* is equivalent to the following

(ii′) $\forall S \in X^{(<\omega)}$ and $t \in X, (u, t) \in \rho$ and $\{v_1, v_2, \cdots, v_k\} \subseteq \rho(S) \Rightarrow t \in \rho(S)$.

(2) $\rho : X \rightharpoonup X$ is finitely regular $\Leftrightarrow \forall (x, y) \in \rho, \exists u \in X$ and $V \in X^{(<\omega)}$ such that

1° $(u, y) \in \rho, V \subseteq \rho(x)$, and

2° $\forall S \in X^{(<\omega)}, t \in X, (u, t) \in \rho$ and $V \subseteq \rho(S) \Rightarrow t \in \rho(S)$.

Definition 3.3. Let ρ be a binary relation on a set X. Define a binary relation $\rho^{(<\omega)}$ on $X^{(<\omega)}$, called the *finite extension* of ρ, by

$$\forall F, G \in X^{(<\omega)}, (F, G) \in \rho^{(<\omega)} \Leftrightarrow G \subseteq \rho(F).$$

Now we give the finitely regular representation and the intrinsic characterization (condition (3) in Theorem 3.2) of hyper-continuous lattices.

Theorem 3.1. *For a binary relation ρ on a set X, the following conditions are equivalent:*

(1) *ρ is finitely regular;*

(2) *the finite extension $\rho^{(<\omega)}$ of ρ is regular;*

(3) *for all $F, G \in X^{(<\omega)}, G \subseteq \rho(F) \Rightarrow \exists U, V \in X^{(<\omega)}$ such that*

(i) *$G \subseteq \rho(U), V \subseteq \rho(F)$, and*

(ii) *for all $S, T \in X^{(<\omega)}, V \subseteq \rho(S)$ and $T \subseteq \rho(U) \Rightarrow T \subseteq \rho(S)$;*

(4) *$(\Phi_\rho(X), \subseteq)$ is a hyper-continuous lattice;*

(5) *$(\{\bigcup_{F \in \mathcal{F}} \rho(F)^{(<\omega)} : \mathcal{F} \subseteq X^{(<\omega)}\}, \subseteq)$ is a completely distributive lattice.*

Proof. (1) ⇒ (2): Define a relation $\delta : X^{(<\omega)} \rightharpoonup X^{(<\omega)}$ by

$$(G,F) \in \delta \Leftrightarrow \forall S,T \in X^{(<\omega)}, G \subseteq \rho(S) \text{ and } F \cap \rho^{-1}(T) \neq \emptyset \Rightarrow T \cap \rho(S) \neq \emptyset. \quad (\dagger)$$

Then

(a) $\rho^{(<\omega)} \circ \delta \circ \rho^{(<\omega)} \subseteq \rho^{(<\omega)}$.

For $H, W \in X^{(<\omega)}$, if $(H,W) \in \rho^{(<\omega)} \circ \delta \circ \rho^{(<\omega)}$, then $\exists G, F \in X^{(<\omega)}$ with $(H,G) \in \rho^{(<\omega)}, ((G,F) \in \delta$ and $(F,W) \in \rho^{(<\omega)}$, i.e., $G \subseteq \rho(H), (G,F) \in \delta$ and $W \subseteq \rho(F)$. For $w \in W$, let $S = H, T = \{w\}$. Then $G \subseteq \rho(S), F \cap \rho^{-1}(T) \neq \emptyset$ (since $w \in \rho(F)$). By $(G,F) \in \delta$ and (†), $T \cap \rho(S) \neq \emptyset$, i.e., $w \in \rho(S) = \rho(H)$. Hence $W \subseteq \rho(H)$. Thus $(H,W) \in \rho^{(<\omega)}$.

(b) $\rho^{(<\omega)} \subseteq \rho^{(<\omega)} \circ \delta \circ \rho^{(<\omega)}$.

For $H, W \in X^{(<\omega)}$, if $(H,W) \in \rho^{(<\omega)}$ (i.e., $W \subseteq \rho(H)$), then $\forall w \in W, \exists h(w) \in H$ with $(h(w), w) \in \rho$. By the finite regularity of ρ, $\exists u(w) \in X$ and $V(w) \in X^{(<\omega)}$ such that

1° $(u(w), w) \in \rho, V(w) \subseteq \rho(h(w))$, and

2° $\forall S \in X^{(<\omega)}, t \in X, (u(w), t) \in \rho$ and $V(w) \subseteq \rho(S) \Rightarrow t \in \rho(S)$.

Let $F = \{u(w) : w \in W\}$ and $G = \bigcup_{w \in W} V(w)$. Then $G \subseteq \bigcup_{w \in W} \rho(h(w)) \subseteq \rho(H)$ and $W \subseteq \rho(F)$, i.e., $(H,W) \in \rho^{(<\omega)}$ and $(F,W) \in \rho^{(<\omega)}$. Now we show that $(G,F) \in \delta$. $\forall S,T \in X^{(<\omega)}$, if $G \subseteq \rho(S)$ and $F \cap \rho^{-1}(T) \neq \emptyset$, then $V(w) \subseteq \rho(S)$ for all $w \in W$ and $\exists (w_0, t_0) \in W \times T$ with $(u(w_0), t_0) \in \rho$. By 2°, $t_0 \in \rho(S)$; and hence $T \cap \rho(S) \neq \emptyset$. By the definition of δ, $(G,H) \in \delta$. It follows from $(H,G) \in \rho^{(<\omega)}$ and $(F,H) \in \rho^{(<\omega)}$ that $(H,W) \in \rho^{(<\omega)} \circ \delta \circ \rho^{(<\omega)}$.

By (a) and (b), $\rho^{(<\omega)} \circ \delta \circ \rho^{(<\omega)} = \rho^{(<\omega)}$. Thus $\rho^{(<\omega)}$ is regular.

(2) ⇒ (3): By the definition of $\rho^{(<\omega)}$ and Theorem 2.2.

(3) ⇒ (4): Let $L = (\Phi_\rho(X), \subseteq)$. For all $M = \rho(A) \in L$, we prove that $M = \cup\{N \in L : N \prec_{\upsilon(L)} M\}$. For any $y \in M$, there is $x \in A$ with $y \in \rho(x)$. By (3), there are $U, V \in X^{(<\omega)}$ such that

(i) $y \in \rho(U), V \subseteq \rho(x)$, and

(ii) For all $S,T \in X^{(<\omega)}, V \subseteq \rho(S)$ and $T \subseteq \rho(U) \Rightarrow T \subseteq \rho(S)$.

Suppose that $V = \{v_1, v_2, \cdots, v_k\}$. For $i \in \{1, 2, \cdots, k\}$, let $N_i = \cup\{N \in L : v_i \notin N\}$ (N_i may be the empty set $\emptyset$). By $V \subseteq \rho(x) \subseteq M$, we have $M \in L \setminus \downarrow \{N_1, N_2, \cdots, N_k\}$. For $N = \rho(B) \in L$, if $N \in L \setminus \downarrow \{N_1, N_2, \cdots, N_k\}$, then by the definition of N_i ($i \in \{1, 2, \cdots, k\}$), $V = \{v_1, v_2, \cdots, v_k\} \subseteq N$. Let $S = B$ and $T = \rho(U)$. Then by (ii), $T = \rho(U) \subseteq \rho(S) = N$. Hence $M \in L \setminus \downarrow \{N_1, N_2, \cdots, N_k\} \subseteq \uparrow \rho(U)$. It follows that $M \in \mathrm{int}_{\upsilon(L)} \uparrow \rho(U)$. Therefore, $y \in \rho(U) \prec_{\upsilon(L)} M$. Whence $M = \cup\{N \in L : N \prec_{\upsilon(L)} M\}$. Thus $L = (\Phi_\rho(X), \subseteq)$ is a hyper-continuous lattice.

(4) $\Rightarrow$ (1): Let $L = (\Phi_\rho(X), \subseteq)$. For all $(x,y) \in \rho$ (i.e., $y \in \rho(x)$), by the hypercontinuity of L, $\exists N = \rho(C) \in L$ with $y \in N \prec_{v(L)} \rho(x)$; and hence $\exists\{M_1, M_2, \cdots, M_n\} \in L^{(<\omega)}$ with $\rho(x) \in L\backslash \downarrow \{M_1, M_2, \cdots, M_n\} \subseteq\uparrow N$. For $i \in \{1,2,\cdots,n\}$, by $\rho(x) \not\subseteq M_i, \exists v_i \in \rho(x)$ with $v_i \notin M_i$. By $y \in \rho(C), \exists u \in C$ with $y \in \rho(u)$. Hence $(u,y) \in \rho, (x,v_i) \in \rho$ $(i = 1,2,\cdots,n)$, i.e., the condition (i) in Definition 3.2 is satisfied. Now we check the condition (ii) in Definition 3.2. For $S = \{s_1, s_2, \cdots, s_n\} \in X^{(<\omega)}$ and $t \in X$, if $(u,t) \in \rho$ and $(s_i, v_i) \in \rho$ for all $i \in \{1,2,\cdots,n\}$, then $\rho(S) = \bigcup\limits_{i=1}^{n} \rho(s_i) \in L\backslash \downarrow \{M_1, M_2, \cdots, M_n\} \subseteq\uparrow N$. Whence $\rho(u) \subseteq N \subseteq \rho(S)$. By $t \in \rho(u), \exists j \in \{1,2,\cdots,n\}$ with $t \in \rho(s_j)$, i.e., $(s_j,t) \in \rho$. Thus $t \in \rho(s_j)$, i.e., $(s_j,t) \in \rho$. Therefore, we get the condition (ii) in Definition 3.2. Whence ρ is finitely regular.

(2) $\Leftrightarrow$ (5): Let $\sigma = \rho^{(<\omega)}$. Then by Theorem 2.1, σ is regular $\Leftrightarrow$ the complete lattice $(\Phi_\sigma(X^{(<\omega)}, \subseteq) = (\{\bigcup\limits_{F\in\mathcal{F}} \rho^{(<\omega)}(F) : \mathcal{F} \subseteq X^{(<\omega)}\}, \subseteq) = (\{\bigcup\limits_{F\in\mathcal{F}} \rho(F)^{(<\omega)} : \mathcal{F} \subseteq X^{(<\omega)}\}, \subseteq)$ is completely distributive.

Theorem 3.2. *For a complete lattice L, the following conditions are equivalent:*

(1) *L is hyper-continuous;*

(2) *the relation $\not\leq$ on L is finitely regular;*

(3) *the finite extension $\not\leq^{(<\omega)}$ of $\not\leq$ is regular;*

(4) *$(\{\bigcup\limits_{F\in\mathcal{F}}(L\backslash \uparrow \vee F)^{(<\omega)} : \mathcal{F} \subseteq L^{(<\omega)}\}, \subseteq)$ is a completely distributive lattice.*

(5) *$\forall x,y \in L, x \not\leq y \Rightarrow \exists u \in L$ and $\{v_1, v_2, \cdots, v_k\} \in L^{(<\omega)}$ such that*

 (i) *$u \not\leq y, x \not\leq v_i$ for all $i \in \{1,2,\cdots,k\}$, and*

 (ii) *$\forall z \in L, u \leq z$ or $z \leq v_j$ for some $j \in \{1,2,\cdots,k\}$;*

(6) *there is a finitely regular relation $\rho : X \rightharpoonup X$ with $L \cong (\Phi_\rho(X), \subseteq)$.*

Proof. (1) $\Rightarrow$ (2): By Theorem 3.1 and $(\Phi_{\not\leq}(L), \subseteq) = (\{L\backslash \uparrow x : x \in L\}, \subseteq) \cong L$.

(2) $\Leftrightarrow$ (3): By Theorem 3.1.

(3) $\Leftrightarrow$ (4): For all $F \in L^{(<\omega)}$, $\not\leq^{(<\omega)}(F) = \{G \in L^{(<\omega)} : G \subseteq \not\leq (F)\} = \{G \in L^{(<\omega)} : G \subseteq L\backslash \uparrow \vee F\} = (L\backslash \uparrow \vee F)^{(<\omega)}$. Hence by Theorem 3.1, we have (3) $\Leftrightarrow$ (4).

(2) $\Rightarrow$ (5): For each $x,y \in L$, by the finite regularity of $\not\leq$, there exist $u \in L$ and $\{v_1, v_2, \cdots, v_k\} \in L^{(<\omega)}$ such that

(i′) $u \not\leq y, x \not\leq v_i$ $(i = 1, 2, \cdots, k)$, and

(ii′) $\forall \{s_1, s_2, \cdots, s_k\} \in L^{(<\omega)}$, $u \not\leq t$ and $s_i \not\leq v_i$ for all $i \in \{1, 2, \cdots, k\}$ $\Rightarrow \exists j \in \{1, 2, \cdots, k\}$ with $s_j \not\leq t$.

For $z \in L$, let $t = z$ and $s_i = z$ $(i = 1, 2, \cdots, k)$. Then by (ii′), $u \leq t = z$ or $\exists j \in \{1, 2, \cdots, k\}$ with $z = s_j \leq v_j$. Hence u and $\{v_1, v_2, \cdots, v_k\}$ satisfy the conditions (i) and (ii) in (5).

(5) ⇒ (2): For $x, y \in L$, if $x \not\leq y$, then by (4), there exist $u \in L$ and $\{v_1, v_2, \cdots, v_k\} \in L^{(<\omega)}$ which satisfy the conditions (i) and (ii) in (5). For $\{s_1, s_2, \cdots, s_k\} \in L^{(<\omega)}$ and $t \in L$, $u \not\leq t$ and $s_i \not\leq v_i$ for all $i \in \{1, 2, \cdots, k\} \Rightarrow \exists j \in \{1, 2, \cdots, k\}$ with $t \leq v_j$ (by the condition (ii)) and $s_j \not\leq v_j \Rightarrow \exists j \in \{1, 2, \cdots, k\}$ with $s_j \leq t$. Thus the relation $\not\leq$ is finitely regular.

(2) ⇒ (6): Let $X = L$ and $\rho = \not\leq$. Then ρ is finitely regular and $L \cong (\Phi_\rho(X), \subseteq)$.

(6) ⇒ (1): By Theorem 3.1.

Finally we propose the following

Problem 3.3. *Does the class of continuous lattices have an (explicit) relational representation?*

Suppose that the class of continuous lattices has a representation of the relations of (explicit) s_0 type. Since $(\Phi_{\not\leq}(L), \subseteq) = \{L \setminus \uparrow x : x \in L\} \cong L$ for all $L \in$**Com**, we have that a complete lattice L is continuous if and only if the relation $\not\leq$ on L has the property s_0. Hence continuous lattices have an intrinsic characterization. But up to now, we do not get any such intrinsic characterizations. Perhaps this is just about as difficult as Problem 3.3.

References

[Ba80] H.-J. Bandelt, Regularity and complete distributivity, *Semigroup Forum*, **19** (1980), 123-126.

[Ba82] ——, On regularity classes of binary relations, In: *Universal Algebra and Applications*, Banach Center Publications, Vol. 9, PWN-Polish Scientific Publishers, 1982, 329-333.

[COMP] G. Gierz, K. Hofmann, K. Keimel, J. D. Lawson, M. Mislove, and D. Scott, *A Compendium of Continuous Lattices*, Springer-Verlag, 1980.

[GL81] G. Gierz and J. D. Lawson, Generalized continuous and hypercontinuous lattices, *Rocky Mountain J. Math.*, **11** (1981), 271-296.

[GLS83] G. Gierz, J. D. Lawson, and A. Stralka, Quasicontinuous posets, *Houston J. Math.*, **9** (1983), 191-208.

[Ma72] G. Markowsky, Idempotents and product representationns with applications to the semigroup of binary relations, *Semigroup Forum*, **5** (1972), 95-119.

[Ra52] G. N. Raney, Completely distributive complete lattices, *Proc. Amer. Math. Soc.*, **3** (1952), 677-680.

[Ra53] ——, A subdirect-union representation for completely distributive complete lattices, *Proc. Amer. Math. Soc.*, **4** (1953), 518-522.

[Ra60] ——, Tight Galois connections and completely distributive complete lattices, *Trans. Amer. Math. Soc.*, **97** (1960), 418-426.

[Sc76] B. M. Schein, Regular elements of the semigroup of all binary relations, *Semigroup Forum*, **13** (1976), 95-102.

[Xu95] X. Q. Xu, Construction of homomorphisms of M-continuous lattices, *Trans. Amer. Math. Soc.*, **347** (1995), 3167-3175.

[Ya69] J. C. Yang, A theorem on the semigroup of binary relations, *Proc. Amer. Math. Soc.*, **22** (1969), 134-135.

[Za63] K. A. Zareckiĭ, The semigroup of binary relations, *Mat. Sbornik*, **61** (1963), 291-305.

Chapter 4

CONVERGENCE CLASSES AND SPACES OF PARTIAL FUNCTIONS

Anthony Karel Seda
Department of Mathematics
University College Cork
Cork, Ireland
a.seda@ucc.ie

Roland Heinze
Institut für Informatik III
Rheinische Friedrich-Wilhelms-Universität Bonn
Römerstr. 164, 53117 Bonn, Germany
heinze@nikocity.de

Pascal Hitzler*
Artificial Intelligence Institute
Dresden University of Technology
01062 Dresden, Germany
phitzler@inf.tu-dresden.de

Abstract We study the relationship between convergence spaces and convergence classes given by means of both nets and filters, we consider the duality between them and we identify in convergence terms when a convergence space coincides with a convergence class. We examine the basic operators in the Vienna Development Method of formal systems development, namely, extension, glueing, restriction, removal and override, from the perspective of the Logic for Computable Functions. Thus, we examine in detail the Scott continuity, or otherwise, of these operators when viewed

*The third named author acknowledges financial support under grant SC/98/621 from Enterprise Ireland.

GQ Zhang et al. (eds.), Domain Theory, Logic and Computation, 75-115.

as operators on the domain $(X \rightarrow Y)$ of partial functions mapping X into Y. The important override operator is not Scott continuous, and we consider topologies defined by convergence classes which rectify this situation.

Keywords: Nets, filters, convergence spaces, convergence classes, (Scott) domains, Scott, Lawson and Cantor topologies and operations on spaces of partial functions, VDM.

1. Introduction

The uses of topology in studying theoretical aspects of computer science are varied and wide. Many of them are related to domain theory and to programming language semantics and hence, ultimately, to the Scott topology. But other important applications are known, including: digital topology in image processing, the use of ideas from homotopy theory in "no deadlock" proofs in concurrency, and the use of topology in logic programming, to mention a few.

Of course, there are also many ways of specifying topologies, varying from the assignment of families of sets to be taken as open, to the use of (ultra)metrics of various generality, through to order-theoretic means. Yet another familiar method, though not, it appears, widely used in computing is to specify convergence by means of convergence classes. Here, one is concerned with conditions on a family of pairs each consisting of a net in a set together with a point of that set so that the given family generates a topology in which the convergent nets (and their limits) are precisely the members of the given class. As a matter of fact, the Scott topology on a domain has a simple characterization in these terms and we will use this later.

On the other hand, convergence structures (known as convergence spaces) more general than topological spaces have been investigated in [BR01], and elsewhere, as a means of unifying discrete and continuous models of computation, or hybrid systems. In a convergence space, one is specifically given a notion of convergence at each point by means of families of filters, see Section 2 and [BR01, Hec00, Hyl79] for details. The notion of convergence thus specified is much weaker than that which prevails in a topological space, although each convergence space has a topology naturally associated with it. Furthermore, the topology just mentioned has, in turn, an associated convergence class, and herein we call a convergence space *topological* if it coincides with its associated convergence class. The embedding of topological spaces into convergence spaces implied thereby is, of course, strict, see [Hec00].

The motivation for this paper is the examination of certain aspects of convergence in computer science, and is threefold, as follows.

First, it is precisely the convergence properties of the topologies used in [HS99, Sed95] which are most useful in relation to termination and semantical questions in logic programming in the presence of negation. Whilst not originally defined by means of convergence, the convergence properties of these topologies are natural and could indeed have been taken as definitive. This point is examined in some detail in [Sed02] where it is shown that convergence can be taken as a fundamental concept in unifying the procedural point of view and the declarative point of view in the context of logic programming with negation.

The second point concerns the Vienna Development Method (VDM) of formal system specification as expounded particularly in the papers [Hug01, Mac90, Mac01]. VDM is a development method which starts with the formal specification of the system requirements and ends, after a sequence of refinement steps, with the implemented program code. At each refinement step, a number of proof obligations have to be fulfilled which ensure that system requirements are met. In the form of VDM developed in [Mac90, Mac01], denoted by VDM$^\clubsuit$ and termed the Irish School of the VDM, preconditions are used, but not postconditions. Instead of using formal logic to verify postconditions, proof obligation (of system invariants) is carried out constructively using a calculus of operators defined on spaces of partial functions. The calculus aims, of course, at reducing complicated calculations to routine symbol manipulation, especially those calculations concerned with things like domain restriction and removal, extension of functions and, in particular, override of functions (which is an important tool in modelling the process of updating records, file systems etc.).

On the other hand, spaces of partial functions, and operators defined on them, arise as particularly important examples of domains in Scott's well-established, and extensive, Logic for Computable Functions (LCF), see [Pau87], which formalizes an abstract model of computability. Thus, although their aims are rather different, it is of interest to contrast VDM$^\clubsuit$ and LCF to the extent of investigating the operators which arise within VDM$^\clubsuit$ from the point of view of LCF, and specifically to determine their computability, or otherwise, in terms of Scott-continuity, and it is the second main purpose of this article to take some initial steps in carrying out this process. Thus, we intend to study in detail the basic operators arising in VDM$^\clubsuit$ when considered as operators on the domain $(X \to Y)$ of partial functions mapping X to Y, and to examine their continuity principally in relation to the Scott topology. However, it turns out that one of the most important operators, the override, is not Scott continuous, and this fact necessitates the introduction of other topologies, related to the Scott topology, to describe its behaviour. The particular topology we introduce here is in fact the smallest refinement of the Scott and Lawson topologies meeting certain natural conditions, see Proposition 3.17. All this is done by means of convergence classes, and we obtain thereby a simple and natural treatment.

Third, we want to examine more closely the relationship between convergence spaces and topological spaces from the point of view of convergence in an attempt to better understand convergence spaces and their applications to modelling hybrid systems, including applications to spaces of valuations in logic programming, and to spaces of partial functions as in Section 3. Indeed, this paper and [Sed02] are complementary: in [Sed02] spaces of valuations in many-valued logics are considered from the point of view of convergence, and here we focus on spaces of partial functions from the same point of view.

In effect, the paper falls naturally into two parts. In the first of these, Section 2, we present a definition of convergence spaces in terms of nets, and a definition of convergence classes in terms of filters, both of which are new. Once that is done, the hoped-for duality between convergence spaces and convergence classes in filter form, on the one hand, and convergence spaces and convergence classes in net form, on the other, can be and, indeed, is established, in Section 2; it is of course derived from the usual duality between nets and filters. Moreover, we view convergence spaces as generalizations of convergence classes (topological convergence spaces), and we are able to identify precise conditions, in terms of convergence, under which a convergence space is topological. In addition, there is the usual advantage of having both formulations available: nets are intuitive and easy to use to check, say, continuity; filters are preferred when features of the space need to be involved.

The second part of the paper, Sections 3 and 4, are devoted to the study of the operators arising in VDM by means of convergence classes, as already mentioned. Taken together with [Sed02], it gives a detailed treatment, based on the convergence concepts in the first part, of two of the main structures encountered in the theory of computation: spaces of valuations and spaces of partial functions. Moreover, it addresses the question, by analogy with areas of mathematical analysis, of what is a reasonable notion of convergence in spaces of partial functions. This question was in fact one of the original motivations for the paper, and the answers we provide are, we believe, both elegant and interesting quite apart from any applications to VDM.

Acknowledgement We thank an anonymous referee for making some suggestions which clarified certain results in the paper, and for drawing our attention to the need to examine the effectiveness of the operators we discuss. This latter point is something which is already under consideration by the authors in relation to the work of [Hug01, Mac90, Mac01] and will be treated in detail elsewhere, but see the remarks made in Section 5, Conclusions and Further Work.

2. Convergence spaces and convergence classes

2.1. Preliminaries

We assume that the reader is familiar with the basic properties of nets, and we use [Kel75, Wil70] as our general references to this topic and for much of our notation. Thus, a *net* in a set X is a function $S : D \to X$, where $(D, \leq)$ is some directed set. The point $S(n), n \in D$, is often denoted S_n or x_n and we frequently refer to "the net $(S_n)_{n\in D}$" or "the net $(x_n)_{n\in D}$" instead of the net S. If no confusion can arise, we use (S_n) as an abbreviation for $(S_n)_{n\in D}$. A *tail* of a net (x_n) in X is a set of the form $\{x_n \,|\, n \geq m\}$, where m is an element of D. As usual, if (x_n) is a net in X, then a property will be said to hold *eventually* with respect to (x_n) if it holds for all $n \geq m$ for some element m of the index set of (x_n).

One point on which we will be specific, however, is in our use of the term "subnet", and we will adopt Kelley's definition throughout (see [Kel75]), noting that this form is more general than that employed in [Wil70]. Thus, a *subnet* T of a net $S : D \to X$ is a net $T : M \to X$ satisfying: (i) $T = S \circ \varphi$, where φ is a function mapping M into D, and (ii) for each $n \in D$, there exists $m \in M$ such that $\varphi(p) \geq n$ whenever $p \geq m$. The point $S \circ \varphi\,(m)$ is often denoted S_{n_m} or x_{n_m}, and we refer to the subnet $(x_{n_m})_{m\in M}$ of $(x_n)_{n\in D}$.

As usual, a net (S_n) in a topological space X will be said to *converge* to $x \in X$, written $S_n \to x$ or $\lim_n S_n = x$, if each neighbourhood U of x contains a tail of (S_n). The following elementary fact will be used quite often in the sequel: *if E is a subset of a topological space X, then $x \in \overline{E}$ if and only if there exists a net (x_n) in E with $x_n \to x$*, where $\overline{E}$ denotes the closure of E in X.

Concerning filters, filter bases and ultrafilters, we again assume a basic familiarity with these topics and refer the reader to [Wil70] for all the background and notation we need. Thus, a filter $\mathcal{A}$ on a set X is a non-empty collection of non-empty sets closed under the processes of taking finite intersections and superset. In particular, a filter $\mathcal{A}$ in a topological space X will be said to *converge* to $x \in X$, written $\mathcal{A} \to x$, if $\mathcal{A}$ is finer than the neighbourhood filter $N(x)$ at x, that is, $N(x) \subseteq \mathcal{A}$. Also, if $A \subseteq X$, then the filter determined by A, namely the set of all supersets of A, will be denoted by $[A]$; in case A is a singleton set $\{x\}$, we denote this ultrafilter by $[x]$ and refer to it as the *point filter at* x. The analogue in terms of filters of the earlier-mentioned elementary fact concerning closure, which again will be used quite often, is the following: *if* E *is a subset of a topological space* X*, then* $x \in \overline{E}$ *if and only if there exists a filter* $\mathcal{A}$ *on* X *with* $E \in \mathcal{A}$ *and* $\mathcal{A} \to x$.

Of course, the theory of nets and the theory of filters are dual in that any topological fact that can be established by means of the one can equally well be established by means of the other. The exact translation of each of these theories into the other can be found in many places, but we follow [Wil70] in this regard and include next the bare details which we will need later.

Let (x_n) be a net in X and, for each $n \in D$, let $B_n = \{x_m \mid m \geq n\}$ be a tail of (x_n). Let $\mathcal{C}$ denote the collection $\{B_n \mid n \in D\}$. Then $\mathcal{C}$ is the base for a filter called the *filter generated by the net* (x_n). On the other hand, let $\mathcal{A}$ be a filter on X and let $D_{\mathcal{A}}$ denote the set $\{(x, A) \mid x \in A \in \mathcal{A}\}$. We define an ordering $\leq$ on $D_{\mathcal{A}}$ by $(x_1, A_1) \leq (x_2, A_2)$ if $A_2 \subseteq A_1$. Then $(D_{\mathcal{A}}, \leq)$ is a directed set, and the mapping $S : D_{\mathcal{A}} \to X$ defined by $S(x, A) = x$ determines a net in X called the *net based on the filter* $\mathcal{A}$. The precise connection between these two notions is contained in the following result.

Theorem 2.1. *Let* X *be a topological space and let* x *be an element of* X*. Then the following statements hold.*
(a) A filter $\mathcal{A}$ *on* X *converges to* x *if and only if the net based on* $\mathcal{A}$ *converges to* x.
(b) A net (x_n) *in* X *converges to* x *if and only if the filter generated by* (x_n) *converges to* x.

Finally, we remind the reader that a closure operator (also known as a Kuratowski, or topological, closure operator) on a set X is a mapping ${}^c : \mathcal{P}(X) \to \mathcal{P}(X)$, from the power set $\mathcal{P}(X)$ of X into itself, subject to the following axioms. (a) $\emptyset^c = \emptyset$. (b) $A \subseteq A^c$ for all $A \subseteq X$. (c) $(A \cup B)^c = A^c \cup B^c$ for all $A, B \subseteq X$. (d) $A^c = (A^c)^c$ for all $A \subseteq X$.

The main fact we need concerning closure operators is the following well-known theorem. *Let* X *be a non-empty set and let* ${}^c : \mathcal{P}(X) \to \mathcal{P}(X)$ *be a closure operator on* X*. Then* $\mathcal{T} = \{X \setminus A \mid A \subseteq X, A = A^c\}$ *is a topology on* X *in which* $\overline{A} = A^c$ *for each subset* A *of* X*. Thus,* A^c *is the topological closure in* X *of each subset* A *of* X *with respect to the topology* $\mathcal{T}$ *determined by* ^{c}.

2.2. Convergence Spaces in Filter and Net Form

We expound here only that part of the theory of convergence spaces which is closely related to the theory of convergence classes as needed later on. For more information on convergence spaces given in terms of filters, see [BR01, Hec00, Hyl79].

We begin this section by studying convergence spaces given in the conventional way in terms of filters. We shall henceforth usually refer to these as "convergence spaces in filter form" in order to distinguish them from the form we give later using nets, which is new. The three results we present concerning convergence spaces in filter form are well-known, see [BR01], but we include those details of proof, of the third, that we need later on.

Definition 2.1 (Convergence Space in Filter Form). Let X be a non-empty set. The pair $(X, \mathcal{F}) = (X, (\mathcal{F}_x)_{x \in X})$ is called a *convergence space in filter form* if, for each $x \in X$, $\mathcal{F}_x$ is a collection of filters on X with the following properties.

(a) The point filter $[x]$ belongs to $\mathcal{F}_x$ for each $x \in X$ (point filter axiom).

(b) If $\mathcal{A} \in \mathcal{F}_x$ and $\mathcal{B} \supseteq \mathcal{A}$ is a filter on X, then $\mathcal{B} \in \mathcal{F}_x$ (closure under superfilters).

One sometimes uses another notation for convergence spaces in filter form. One writes $\mathcal{A} \downarrow x$ if and only if $\mathcal{A} \in \mathcal{F}_x$ and refers to the convergence space $(X, \downarrow)$ instead of $(X, \mathcal{F})$. Thus, we interpret $\downarrow$ as a relation between filters on X and elements of X. If $\mathcal{A} \in \mathcal{F}_x$, we say $\mathcal{A}$ *converges to* x.

We say that a convergence space $(X, \mathcal{F})$ in filter form is *pointed* if, for each $x \in X$, we have $\bigcap \mathcal{F}_x \in \mathcal{F}_x$.

Finally, a subset $O \subseteq X$ is said to be *open* in the *induced topology* of a convergence space $(X, \mathcal{F})$ in filter form if $\mathcal{A} \downarrow x \in O$ always implies $O \in \mathcal{A}$ (so that $\mathcal{A} \supseteq [O]$), where $\mathcal{A} \downarrow x \in O$ means "$\mathcal{A}$ converges to x and $x \in O$".

Lemma 2.2. *For each $x \in X$, $\bigcap \mathcal{F}_x$ is a filter coarser than each $\mathcal{A} \in \mathcal{F}_x$. Furthermore, for each $A \in \bigcap \mathcal{F}_x$, we have $x \in A$.*

Lemma 2.3. *The induced topology of a convergence space $(X, \mathcal{F})$ in filter form is a topology on X.*

Lemma 2.4. *Every topological space $(X, \mathcal{T})$ is representable as a convergence space $(X, \mathcal{F})$ in filter form such that the induced topology of $(X, \mathcal{F})$ is $\mathcal{T}$.*

Proof. For each $x \in X$, let $\mathcal{F}_x$ be the set of all filters $\mathcal{A}$ on X such that $\mathcal{A} \supseteq N(x)$, where $N(x)$ denotes the neighbourhood filter at x with respect to the topology $\mathcal{T}$. Obviously, $\mathcal{F}_x$ has the properties listed in the definition of a convergence space in filter form. So, we choose $\mathcal{F} = (\mathcal{F}_x)_{x \in X}$.

First, let $O \in \mathcal{T}$ and suppose that $\mathcal{A} \downarrow x \in O$. We immediately get $O \in N(x) \subseteq \mathcal{A}$ and, as x was chosen arbitrarily, we see that O is open with respect to the induced topology.

Next, let O be open with respect to the induced topology. Then $N(x) \downarrow x \in O$ implies $O \in N(x)$. Therefore, there exists an open set $O_x \in N(x) \cap \mathcal{T}$ with $O_x \subseteq O$, and we obtain $O = \bigcup_{x \in O} O_x \in \mathcal{T}$. So, O is open with respect to $\mathcal{T}$, as required. □

Definition 2.2. Let $(X, \mathcal{T})$ be a topological space. The *convergence space in filter form induced by* $\mathcal{T}$ is defined as follows: $\mathcal{A} \downarrow_{\mathcal{O}} x$ if and only if $\mathcal{A} \supseteq N(x)$ is a filter on X, where $N(x)$ is the neighbourhood filter at x in $\mathcal{T}$. A convergence space $(X, \downarrow)$ in filter form is called a *topological convergence space (in filter form)* if there is a topology $\mathcal{T}$ on X with the property that the convergence space in filter form induced by $\mathcal{T}$ coincides with $(X, \downarrow)$. We sometimes use the notation $(X, \downarrow_{\mathcal{O}})$ to indicate that a convergence space is a topological convergence space.

We now turn our attention to formulating the notion of convergence space in terms of nets, and obtain duals of each of the results above. Results of the later parts of this section show that the definition we give and the corresponding duality are both highly satisfactory.

Definition 2.3 (Convergence Space in Net Form). Let X be a non-empty set. The pair $(X, \mathcal{S}) = (X, (\mathcal{S}_s)_{s \in X})$ is called a *convergence space in net form* if, for each $s \in X$, $\mathcal{S}_s$ is a non-empty collection of nets in X with the following properties.

(a) If $S \colon D \to X$ is a constant net, that is, $S_n = s \in X$ for all $n \in D$, then $S \in \mathcal{S}_s$.

(b) If $S \in \mathcal{S}_s$ and T is a subnet of S, then $T \in \mathcal{S}_s$.

We will sometimes use another notation for convergence spaces in net form. We write $S \downarrow s$ if and only if $S \in \mathcal{S}_s$ and refer to the convergence space $(X, \downarrow)$ instead of $(X, \mathcal{S})$. Thus, we interpret $\downarrow$ as a relation between nets in X and elements of X. If $S \in \mathcal{S}_s$, we say S *converges to* s.

A subset $O \subseteq X$ is said to be *open* in the *induced topology* of a convergence space $(X, \mathcal{S})$ in net form if $S \downarrow s \in O$ always implies that there exists $n \in D$ with $S_m \in O$ for all $m \geq n$. Here, of course, $S \downarrow s \in O$ means "S converges to s and $s \in O$".

By analogy with the results above for convergence spaces in filter form, we prove that the induced topology just defined really is a topology, and that every topological space is representable as a convergence space in net form in such a way that the induced topology of the convergence space coincides with the original topology of the topological space.

Lemma 2.5. *The induced topology of a convergence space* $(X, \mathcal{S})$ *in net form is a topology on* X.

Proof. From the definition, $\emptyset$ and X are obviously open sets in the induced topology.

Let $(O_i)_{i \in I}$ be a family of open sets in the induced topology of $(X, \mathcal{S})$. Let $O = \bigcup_{i \in I} O_i$, and suppose that $S \downarrow s \in O$. Then there exists $i \in I$ and $n \in D$ such that $s \in O_i$ and $S_m \in O_i \subseteq O$ for all $m \geq n$. Since $s \in O$ was chosen arbitrarily, we see that O is an element of the induced topology.

Now let $O_1, O_2 \subseteq X$ be open sets in the induced topology of $(X, \mathcal{S})$. Let $O = O_1 \cap O_2$ and suppose that $S \downarrow s \in O$. Then there exist $n_1, n_2 \in D$ such that $S_n \in O_i$ for all $n \geq n_i, i = 1, 2$. Since D is directed, there exists $n_0 \in D$ with $n_0 \geq n_1, n_2$, and we have $S_n \in O$ for all $n \geq n_0$. Because s is an arbitrary element of O, we conclude that O is an element of the induced topology, as required. □

Lemma 2.6. *Every topological space* $(X, \mathcal{T})$ *is representable as a convergence space* $(X, \mathcal{S})$ *in net form such that the induced topology of* $(X, \mathcal{S})$ *is* $\mathcal{T}$.

Proof. We define $S \in \mathcal{S}_s$ if and only if $S \colon D \to X$ is a net with $S_n \to s$ with respect to $\mathcal{T}$. Obviously $\mathcal{S}_s, s \in X$, fulfills the conditions listed in the definition of a convergence space in net form. So, we choose $\mathcal{S} = (\mathcal{S}_s)_{s \in X}$.

We show that the induced topology of $(X, \mathcal{S})$ is $\mathcal{T}$. First, let O be open with respect to $\mathcal{T}$ and suppose that $S \downarrow s \in O$. By definition, we have $S_n \to s$ with respect to $\mathcal{T}$ and, because $O \in \mathcal{T}$, we conclude that there exists $n \in D$ with $S_m \in O$ for all $m \geq n$. Since s is an arbitrary element of O, we see that O is an open set of the induced topology of $(X, \mathcal{S})$.

Next, let O be open with respect to the induced topology. Suppose $O \notin \mathcal{T}$ so that $X \setminus O$ is not closed with respect to $\mathcal{T}$. Then there exists $s \in \overline{X \setminus O}^{\mathcal{T}} \cap O$, where $\overline{X \setminus O}^{\mathcal{T}}$ denotes the closure of $X \setminus O$ in X relative to $\mathcal{T}$. Thus, by the elementary facts noted earlier, there exists a net $S : D \to X$ with $S_n \in X \setminus O$ for all $n \in D$, and also $S_n \to s \in O$. Hence, we have $S \downarrow s \in O$. But O is open in the induced topology. Therefore, there exists $n \in D$ such that $S_m \in O$ for all $m \geq n$, which contradicts the fact that $S_n \in X \setminus O$ for all $n \in D$. So, we conclude that $O \in \mathcal{T}$, as required. □

Definition 2.4. Let $(X, \mathcal{T})$ be a topological space. The *convergence space in net form induced by* $\mathcal{T}$ is defined as follows: $S \downarrow_{\mathcal{O}} s$ if and only if S is a net in X with $S_n \to s$ with respect to $\mathcal{T}$. A convergence space $(X, \downarrow)$ in net form is called a *topological convergence space (in net form)* if there is a topology $\mathcal{T}$ on X with the property that the convergence space in net form induced by $\mathcal{T}$ coincides with $(X, \downarrow)$. We sometimes use the notation $(X, \downarrow_{\mathcal{O}})$ to indicate that a convergence space is a topological convergence space.

2.3. Convergence Classes in Net and Filter Form

As already noted, convergence spaces are normally defined in terms of filters whilst convergence classes are defined in terms of nets. We begin this section by briefly considering convergence classes defined by means of nets, following [Kel75], before presenting a treatment of them defined by means of filters which gives the duality we want between the two approaches. The same terminology (net and filter form) as used in the previous section will be adopted here and in the sequel to distinguish the two definitions.

Definition 2.5 (Convergence Class in Net Form). Let X be an arbitrary non-empty set. We call $\mathcal{C}$ a *convergence class for* X *in net form* if $\mathcal{C}$ is a set of pairs each consisting of a net S in X and a point s of X such that the conditions listed below are satisfied. Instead of $(S, s) \in \mathcal{C}$ we also use the notation S *converges* $(\mathcal{C})$ *to* s or $\lim_n S_n \equiv s\,(\mathcal{C})$, see [Kel75, Page 74].

(a) If $S \colon D \to X$ is a constant net in X, that is, $S_n = s \in X$ for all $n \in D$, then $(S, s) \in \mathcal{C}$.

(b) If $(S, s) \in \mathcal{C}$ and T is a subnet of S, then $(T, s) \in \mathcal{C}$.

(c) If $(S, s) \notin \mathcal{C}$, then there exists a subnet T of S such that for every subnet R of T we have $(R, s) \notin \mathcal{C}$.

(d) Let D be a directed set, let E_m be a directed set for each $m \in D$, let F denote the product directed set $D \times \prod_{m \in D} E_m$, and let F' denote the fibred product $D \times_D \bigcup_{m \in D} E_m = \{(m, n) \mid m \in D, n \in E_m\}$. Let $R : F \to F'$ be defined by $R(m, f) = (m, f(m))$ for each $(m, f) \in F$ and let $S : F' \to X$ be a function. If $\lim_m \lim_n S(m, n) \equiv s(\mathcal{C})$, then $(S \circ R, s) \in \mathcal{C}$.

A few comments concerning this definition are in order. First, conditions (a) and (b) reflect elementary properties of net convergence in a topological space. Second, if a net $S\colon D \to X$ does not converge to s in the topological space X, there must exist $U \in N(s)$ and a cofinal subset $D' \subseteq D$ such that $S_n \in X \setminus U$ for all $n \in D'$. This fact is the reason for stipulating condition (c) in the above definition. Third, the iterated limits theorem, see [Kel75, page 69], is the motivation for condition (d) in the definition. Finally, by a product directed set $\prod_{m\in D} I_m$, we understand of course the pointwise ordering on the product $\prod_{m\in D} I_m$ of the directed sets I_m; thus, for elements f and g of $\prod_{m\in D} I_m$, we have $f \leq g$ if and only if $f(m) \leq g(m)$ for each $m \in D$.

We now record the main theorem concerning convergence classes in net form. This result is basic to the sort of applications we make later in this paper and elsewhere. However, the last part of the proof given in [Kel75, Theorem 9, page 75] appears to be incorrect (the net $\{T \circ U(m,n), n \in E_m\}$ defined there is clearly not defined), and therefore we take the trouble to fill this gap.

Theorem 2.7. *Let $\mathcal{C}$ be a convergence class in net form for a non-empty set X. For each $A \subseteq X$, let A^c denote the set of those $s \in X$ such that there is a net S in A with $(S, s) \in \mathcal{C}$. Then ^{c} is a closure operator on X and hence defines a topology $\mathcal{T}$ on X. Moreover, we have $(S, s) \in \mathcal{C}$ if and only if $S_n \to s$ with respect to $\mathcal{T}$.*

Proof. Following the proof in [Kel75, Theorem 9, pp. 74–75], suppose that it is already established that ^{c} is a closure operator on X, and that convergence ($\mathcal{C}$) of S to s implies convergence $S_n \to s$ relative to $\mathcal{T}$. Then it remains to show that convergence $S_n \to s$ with respect to $\mathcal{T}$ implies that $(S, s) \in \mathcal{C}$. Suppose in fact that $(S, s) \notin \mathcal{C}$. By condition (c) in the definition of $\mathcal{C}$, there exists a subnet $T : D \to X$ of S such that for each subnet R of T we have $(R, s) \notin \mathcal{C}$. For each $m \in D$, let $D_m = \{n \in D \mid n \geq m\}$ and let $A_m = T(D_m)$. Since D_m is cofinal in D, we have that $T|_{D_m}$ is a subnet of T which must converge to s with respect to $\mathcal{T}$ since S and, hence T, have this property. Using the elementary facts quoted earlier concerning nets and closure, the fact that ^{c} defines the topology $\mathcal{T}$ and the fact that closure relative to ^{c} is the same thing as closure relative to $\mathcal{T}$, we get $s \in (A_m)^c$ for each $m \in D$. Therefore, we obtain, for each $m \in D$, a net $U(m, \bullet) : E_m \to A_m$ with $(U(m, \bullet), s) \in \mathcal{C}$. We apply condition (d) in the definition of $\mathcal{C}$. Let F and R be as defined in condition (d). Then we get $(U \circ R, s) \in \mathcal{C}$. Because we have $U \circ R(m, f) \in A_m$, there exists $n_{m,f} \in D_m$ with $U \circ R(m, f) = T_{n_{m,f}}$ for all $(m, f) \in F$. We define $\varphi : F \to D$ by $\varphi(m, f) = n_{m,f}$ for all $(m, f) \in F$ and obtain $U \circ R = T \circ \varphi$. Finally, given $m \in D$, take any

$(m, f) \in F$, that is, choose any f. Then, if $(m', g) \geq (m, f)$, we have $\varphi(m', g) = n_{m',g} \geq m' \geq m$. Therefore, $U \circ R$ is a subnet of T and $(U \circ R, s) \in \mathcal{C}$, which gives a contradiction to our present assumption. We therefore conclude that $(S, s) \in \mathcal{C}$ to finish the proof. □

We now turn to the main topic of this subsection, namely, the provision of conditions on classes of filters which ensure that they determine a topology in which the resulting convergent filters are precisely the filters first given. The first step is to provide a suitable filter form of the theorem on iterated limits for nets, as follows.

Theorem 2.8. *Let D be an index set, let $(\mathcal{F}_d)_{d\in D}$ be a family of filters on a topological space $(X, \mathcal{T})$, let $(A_d)_{d\in D}$ be a family of subsets of X such that $\{A_d \,|\, d \in D\}$ is a filter base on X and let $S = \{s_d \,|\, d \in D\} \subseteq X, s \in X$. Suppose that for all $d \in D$ we have*

$$A_d \in \mathcal{F}_d,\ \mathcal{F}_d \to s_d \text{ and } \forall s' \in S\ \exists\, d' \in D \text{ s.t. } A_d \in \mathcal{F}_{d'},\ \mathcal{F}_{d'} \to s'. \quad (4.1)$$

Let $\mathcal{F}$ be a filter with $S \in \mathcal{F}$ and such that $\mathcal{F} \to s$. Then there exists a filter $\mathcal{G}$ on X with $A_d \in \mathcal{G}$ for all $d \in D$ and $\mathcal{G} \to s$.

Proof. Suppose that the premises of our claim are satisfied. From condition (4.1), we conclude $s_d \in S \subseteq \overline{A_d}$ for all $d \in D$. Because $S \in \mathcal{F}$ and $\mathcal{F} \to s$, we obtain $s \in \overline{S}$. In particular, we have $s \in \overline{S} \subseteq \overline{A_d}$ for all $d \in D$. Let $\mathcal{B} = \{A_d \cap U \,|\, d \in D, U \in N(s)\}$, noting that these sets are non-empty since $s \in \overline{A_d}$ for all d. Then $\mathcal{B}$ is a filter base because, for all $d, d' \in D, U, U' \in N(s)$, there exists $d'' \in D$ with $A_{d''} \subseteq A_d \cap A_{d'}$ and hence

$$\emptyset \neq A_{d''} \cap (U \cap U') \subseteq (A_d \cap A_{d'}) \cap (U \cap U') = (A_d \cap U) \cap (A_{d'} \cap U').$$

Let $\mathcal{G}$ be the filter with base $\mathcal{B}$. We have $N(s) \subseteq \mathcal{G}$ and $A_d \in \mathcal{G}$ for all $d \in D$. In particular, we obtain $\mathcal{G} \to s$. □

Condition (d) in the definition of convergence classes in filter form given below is inspired by Theorem 2.8. Condition (a) is inspired by considering the filter generated by a constant net, and condition (b) by the fact that every filter $\mathcal{G}$ finer than $\mathcal{F}$ with $\mathcal{F}$ converging to x also converges to x. Finally, condition (c) is clearly necessary by elementary properties of convergence. A similar definition of classes in filter form can be found in [Smy92] where conditions (a) to (c) only of the following definition are used, but with correspondingly weaker conclusions.

Definition 2.6 (Convergence Class in Filter Form). Let X be a non-empty set. A *convergence class $\mathcal{C}$ for X in filter form* is a set of pairs $(\mathcal{F}, s)$ each consisting of a filter $\mathcal{F}$ on X and an element s of X subject to the following conditions. If $(\mathcal{F}, s) \in \mathcal{C}$, we say $\mathcal{F}$ *converges* $(\mathcal{C})$ *to* s and sometimes write $\mathcal{F} \to s\,(\mathcal{C})$.

(a) Let $s \in X$ and let $\mathcal{F} = \{F \subseteq X \,|\, s \in F\} = [s]$ be the point ultrafilter on X at s. Then $(\mathcal{F}, s) \in \mathcal{C}$.

(b) If $(\mathcal{F}, s) \in \mathcal{C}$ and $\mathcal{G}$ is a filter on X finer than $\mathcal{F}$, then $(\mathcal{G}, s) \in \mathcal{C}$.

(c) If $(\mathcal{F}, s) \notin \mathcal{C}$, then there exists a filter $\mathcal{F}' \supseteq \mathcal{F}$ such that, for each filter $\mathcal{G} \supseteq \mathcal{F}'$, we have $(\mathcal{G}, s) \notin \mathcal{C}$.

(d) Let D be an index set, let $(\mathcal{F}_d)_{d \in D}$ be a family of filters on X, let $(A_d)_{d \in D}$ be a family of subsets of X such that $\{A_d \,|\, d \in D\}$ is a filter base on X and let $S = \{s_d \,|\, d \in D\} \subseteq X, s \in X$. Suppose that for all $d \in D$ we have the following condition holding

$A_d \in \mathcal{F}_d$, $(\mathcal{F}_d, s_d) \in \mathcal{C}$ and $\forall\, s' \in S\, \exists\, d' \in D$ such that $A_d \in \mathcal{F}_{d'}$, $(\mathcal{F}_{d'}, s') \in \mathcal{C}$.

Let $\mathcal{F}$ be a filter with $S \in \mathcal{F}$ and such that $(\mathcal{F}, s) \in \mathcal{C}$. Then there exists a filter $\mathcal{G}$ on X with $A_d \in \mathcal{G}$ for all $d \in D$ and $(\mathcal{G}, s) \in \mathcal{C}$.

The main theorem concerning convergence classes in filter form is the following analogue of Theorem 2.7.

Theorem 2.9. *Let $\mathcal{C}$ be a convergence class in filter form for a non-empty set X. For each $A \subseteq X$, let A^c denote the set of those $s \in X$ such that there is a filter $\mathcal{F}$ on X with $A \in \mathcal{F}$ and $(\mathcal{F}, s) \in \mathcal{C}$. Then ^{c} is a closure operator on X and hence defines a topology $\mathcal{T}$ on X. Moreover, we have $(\mathcal{F}, s) \in \mathcal{C}$ if and only if $\mathcal{F} \to s$ with respect to $\mathcal{T}$.*

Proof. First we show that ^{c} is a closure operator on X.
(i) It is clear that $\emptyset^c = \emptyset$.
(ii) We show $A \subseteq A^c$. Let $s \in A$. We have $A \in \mathcal{F} = \{F \subseteq X \,|\, s \in F\} = [s]$. Using condition (a) in the definition of $\mathcal{C}$, we get $(\mathcal{F}, s) \in \mathcal{C}$ and conclude that $s \in A^c$.
(iii) We show $(A \cup B)^c = A^c \cup B^c$. Let $s \in A^c$. Then there exists a filter $\mathcal{F}$ on X with $A \in \mathcal{F}$ and $(\mathcal{F}, s) \in \mathcal{C}$. Since $\mathcal{F}$ is a filter, we obtain that $(A \cup B) \in \mathcal{F}$ and conclude that $s \in (A \cup B)^c$. In the same way, one proves $B^c \subseteq (A \cup B)^c$. Now let $s \in (A \cup B)^c$. Then there exists a filter $\mathcal{F}$ on X with $(A \cup B) \in \mathcal{F}$ and $(\mathcal{F}, s) \in \mathcal{C}$. Let $\mathcal{B}_1 = \{F \cap A \,|\, F \in \mathcal{F}\}$ and $\mathcal{B}_2 = \{F \cap B \,|\, F \in \mathcal{F}\}$. Assume $\emptyset \in \mathcal{B}_1 \cap \mathcal{B}_2$. We obtain $F_1, F_2 \in \mathcal{F}$ with $F_1 \cap A = F_2 \cap B = \emptyset$ and conclude $(F_1 \cap F_2) \cap (A \cup B) = \emptyset$, which contradicts the fact that $(A \cup B), (F_1 \cap F_2) \in \mathcal{F}$. Therefore, $\mathcal{B}_1$ or $\mathcal{B}_2$ is a base for a filter $\mathcal{F}' \supseteq \mathcal{F}$ with $A \in \mathcal{F}'$ or $B \in \mathcal{F}'$. Using condition (b) in the definition of $\mathcal{C}$, we conclude that $(\mathcal{F}', s) \in \mathcal{C}$. So, $s \in A^c \cup B^c$.

(iv) We show $(A^c)^c = A^c$. Using (ii), we have $A^c \subseteq (A^c)^c$. Let $s \in (A^c)^c$. Then there exists a filter $\mathcal{F}$ with $A^c \in \mathcal{F}$ and $(\mathcal{F}, s) \in \mathcal{C}$. For each $a \in A^c$, there exists a filter $\mathcal{F}_a$ with $A \in \mathcal{F}_a$ and $(\mathcal{F}_a, a) \in \mathcal{C}$. We use condition (d) in the definition of $\mathcal{C}$. Let $D = A^c$, let $A_a = A$ and $s_a = a$ for all $a \in D$, and let $S = A^c$. Then the premises of condition (d) are satisfied, and there exists a filter $\mathcal{G}$ on X with $A \in \mathcal{G}$ and $(\mathcal{G}, s) \in \mathcal{C}$. We conclude that $s \in A^c$.

Next we prove the equivalence statement in the theorem.
(v) We show $(\mathcal{F}, s) \in \mathcal{C}$ implies $\mathcal{F} \to s$ with respect to $\mathcal{T}$. Suppose that $\mathcal{F} \to s$ with respect to $\mathcal{T}$ is false. Then there exists $U \in N(s) \cap \mathcal{T}$ with $U \notin \mathcal{F}$. Let $\mathcal{B} = \{F \cap (X \setminus U) \,|\, F \in \mathcal{F}\}$. Since, for all $F \in \mathcal{F}$, $F \cap (X \setminus U) \neq \emptyset$ (otherwise there would exist $F \in \mathcal{F}$ with $F \subseteq U$, that is, $U \in \mathcal{F}$, which is a contradiction), $\mathcal{B}$ is a base for a filter $\mathcal{F}' \supseteq \mathcal{F}$. Using condition (b) in the definition of $\mathcal{C}$, it follows that $(\mathcal{F}', s) \in \mathcal{C}$. Because $B \subseteq X \setminus U$ for all $B \in \mathcal{B}$, we have $X \setminus U \in \mathcal{F}'$ and so we get $s \in (X \setminus U)^c$ in contradiction to $X \setminus U = (X \setminus U)^c$ and $s \in U$. So, $\mathcal{F} \to s$ with respect to $\mathcal{T}$ is true.
(vi) We show that $\mathcal{F} \to s$ with respect to $\mathcal{T}$ implies that $(\mathcal{F}, s) \in \mathcal{C}$. Suppose $(\mathcal{F}, s) \notin \mathcal{C}$. By condition (c) in the definition of $\mathcal{C}$, there exists a filter $\mathcal{F}' \supseteq \mathcal{F}$ such that, for all filters $\mathcal{G} \supseteq \mathcal{F}'$, we have $(\mathcal{G}, s) \notin \mathcal{C}$. We have $N(s) \subseteq \mathcal{F} \subseteq \mathcal{F}'$. Therefore, we obtain $s \in \bigcap_{F \in \mathcal{F}'} \overline{F}$. The definition of ^{c} and the equality of closure in ^{c} and in $\mathcal{T}$ yields, for each $F \in \mathcal{F}'$, a filter $\mathcal{F}_F$ with $F \in \mathcal{F}_F$ and $(\mathcal{F}_F, s) \in \mathcal{C}$. We use condition (d) in the definition of $\mathcal{C}$. Let $D = \mathcal{F}'$, let $A_F = F$, and let $s_F = s$ for each $F \in \mathcal{F}'$. Due to condition (a) in the definition of $\mathcal{C}$, we get, for the point ultrafilter $\mathcal{F}'' = [s]$, the property $(\mathcal{F}'', s) \in \mathcal{C}$. So, the premises of condition (d) are satisfied and we obtain a filter $\mathcal{G}$ on X with $F \in \mathcal{G}$ for all $F \in \mathcal{F}'$ and $(\mathcal{G}, s) \in \mathcal{C}$. In particular, $\mathcal{G} \supseteq \mathcal{F}'$ which is a contradiction to our present assumption. Thus, we conclude $(\mathcal{F}, s) \in \mathcal{C}$, as required to finish the proof. □

Remark 2.1. It is an immediate consequence of the definitions that each convergence class in net form resp. filter form is a convergence space in net form resp. filter form.

Some natural questions now immediately arise as a consequence of the results above.
(1) Every convergence class induces a topology on the underlying space. This topology induces on the other hand a convergence space (see Lemma 2.4 and Lemma 2.6). Is this convergence space once again a convergence class in net or filter form? If so, is this convergence class equal to the original convergence class?
(2) Is every convergence class in filter form a pointed convergence space?

Is the induced topology of the convergence space equal to the induced topology of the convergence class?
(3) Can one transform each convergence class in net form into a convergence class in filter form (and vice versa) such that both induce the same topology?

It is the purpose of the rest of this section to give positive answers to all these questions, and we proceed to do this next. In the following, we denote the induced topology of a convergence space $(X, \mathcal{S})$ or $(X, \mathcal{F})$ by TX. Each convergence class will also be interpreted as a convergence space with extra properties. Therefore, if we speak of a convergence class we will sometimes use the notation employed for convergence spaces to denote elements of the convergence class.

2.4. Properties of Convergence Classes in Net Form

Lemma 2.10. *Let X be a non-empty set and let $\mathcal{C}$ be a convergence class for X in net form. Then the induced topology $\mathcal{T}$ of the convergence class coincides with TX, that is, $TX = \mathcal{T}$.*

Proof. Let $O \in TX$ be an open set of the underlying convergence space in net form. We show that $X \setminus O$ is closed with respect to $\mathcal{T}$. Suppose that there exists $s \in (X \setminus O)^c \cap O$. Then there exists a net $S : D \to X \setminus O$ with $S_n \downarrow s \in O$. Since $O \in TX$, there exists $n \in D$ such that $S_m \in O$ for all $m \geq n$, which contradicts the fact that $S_n \in X \setminus O$ for all $n \in D$. We conclude $(X \setminus O)^c = X \setminus O$ or, in other words, that $O \in \mathcal{T}$.

Now let $O \in \mathcal{T}$ and suppose that $S \downarrow s \in O$, so that $S : D \to X$ is a net and $S_n \to s \in O$ with respect to $\mathcal{T}$. We conclude that there exists $n \in D$ with $S_m \in O$ for all $m \geq n$. Since $s \in O$ was chosen arbitrarily and because of the definition of TX, we get $O \in TX$. □

Lemma 2.11 (Associated Convergence Space). *Let X be a non-empty set and let $\mathcal{C}$ be a convergence class for X in net form. Let $\mathcal{T}$ be the induced topology of $\mathcal{C}$, let $(X, \downarrow_{\mathcal{O}})$ be the induced convergence space (see Definition 2.4) with respect to the topology $\mathcal{T}$ and, finally, let $S \downarrow_{\mathcal{C}} s$ if and only if $(S, s) \in \mathcal{C}$. Then we have $\downarrow_{\mathcal{O}} = \downarrow_{\mathcal{C}}$.*

Proof. Applying Definition 2.4 and using Theorem 2.7 we conclude, for each net S in X and $s \in X$, that

$$S \downarrow_{\mathcal{O}} s \iff S_n \to s \text{ with respect to } \mathcal{T} \iff S \downarrow_{\mathcal{C}} s.$$

□

Lemma 2.12. *Let $(X, \mathcal{T})$ be a topological space. Let $(X, \downarrow_{\mathcal{O}})$ be the induced convergence space in net form with respect to the topology $\mathcal{T}$. We define $(S, s) \in \mathcal{C}$ if and only if $S \downarrow_{\mathcal{O}} s$. Then $\mathcal{C}$ is a convergence class in net form.*

Proof. Since $(X, \downarrow_{\mathcal{O}})$ is a convergence space in net form, conditions (a) and (b) in the definition of convergence class in net form are already satisfied. We have to verify conditions (c) and (d). We have

$$(S, s) \in \mathcal{C} \Longleftrightarrow S \downarrow_{\mathcal{O}} s \Longleftrightarrow S_n \to s \text{ with respect to } \mathcal{T}.$$

First we prove condition (c). Let $S : D \to X$ be a net and let $(S, s) \notin \mathcal{C}$, so that (S_n) does not converge to s relative to $\mathcal{T}$. Then there is a neighbourhood U of s, and a cofinal subset D' of D such that $S_m \in X \setminus U$ for all $m \in D'$. Then the restriction of S to D' is a subnet T of S with the property that every subnet R of T fails to converge to s relative to $\mathcal{T}$. So, for each subnet R of T, we have $(R, s) \notin \mathcal{C}$, as required.

Condition (d) follows immediately because we defined $(S, s) \in \mathcal{C}$ if and only if $S_n \to s$ with respect to $\mathcal{T}$. We have only to apply the theorem on iterated limits ([Kel75, Page 69]) to finish the argument. □

2.5. Properties of Convergence Classes in Filter Form

Lemma 2.13. *Let X be a non-empty set and let $\mathcal{C}$ be a convergence class for X in filter form. Then the induced topology $\mathcal{T}$ of the convergence class coincides with TX, that is, $TX = \mathcal{T}$.*

Proof. Let $O \in TX$ be an open set of the underlying convergence space in filter form. We show that $X \setminus O$ is closed with respect to $\mathcal{T}$. Assume that there exists $s \in (X \setminus O)^c \cap O$. Then there exists a filter $\mathcal{F}$ on X with $X \setminus O \in \mathcal{F}$ and $\mathcal{F} \downarrow s \in O$. Since $O \in TX$, we conclude that $O \in \mathcal{F}$, which contradicts $X \setminus O \in \mathcal{F}$. Thus, we get $X \setminus O = (X \setminus O)^c$, so that O is open with respect to $\mathcal{T}$.

Now let $O \in \mathcal{T}, O \neq \emptyset$. Let $\mathcal{A}$ be a filter on X with $\mathcal{A} \downarrow x \in O$. Assume that $O \notin \mathcal{A}$. Then, for all $B \subseteq O$, we have $B \notin \mathcal{A}$, and we conclude that for all $B \in \mathcal{A}$ we have $B \cap (X \setminus O) \neq \emptyset$. We define $\mathcal{B} = \{B \cap (X \setminus O) \,|\, B \in \mathcal{A}\}$. Then $\mathcal{B}$ is a base for a filter $\mathcal{G} \supseteq \mathcal{A}$. Therefore, we obtain $\mathcal{G} \downarrow x \in O$ and $X \setminus O \in \mathcal{G}$. This means that $x \in (X \setminus O)^c = X \setminus O$ as $O \in \mathcal{T}$, and we get a contradiction. Thus, $O \in \mathcal{A}$ and, as $x \in O$ is arbitrary, we conclude that $O \in TX$. □

Lemma 2.14 (Associated Convergence Space). *Let X be a non-empty set and let $\mathcal{C}$ be a convergence class for X in filter form. Let $\mathcal{T}$ be the induced topology of $\mathcal{C}$, let $(X, \downarrow_{\mathcal{O}})$ be the induced convergence space* *(see Definition 2.2) with respect to the topology $\mathcal{T}$ and, finally, let $\mathcal{F} \downarrow_{\mathcal{C}} x$ if and only if $(\mathcal{F}, x) \in \mathcal{C}$. Then we have $\downarrow_{\mathcal{O}} = \downarrow_{\mathcal{C}}$.*

Proof. Applying Definition 2.2 and using Theorem 2.9 we conclude, for each filter $\mathcal{F}$ on X and $x \in X$, that

$$\mathcal{F} \downarrow_{\mathcal{O}} x \Longleftrightarrow \mathcal{F} \to x \text{ with respect to } \mathcal{T} \Longleftrightarrow \mathcal{F} \downarrow_{\mathcal{C}} x.$$

□

Corollary 2.15. *Let X be a non-empty set. Then every convergence class $\mathcal{C}$ for X in filter form is a pointed convergence space in filter form.*

Proof. Let $x \in X$. We have $\bigcap \mathcal{F}_x = \bigcap \{\mathcal{A} \,|\, \mathcal{A} \downarrow_{\mathcal{C}} x\} = \bigcap \{\mathcal{A} \,|\, \mathcal{A} \downarrow_{\mathcal{O}} x\}$, and the definition of an induced convergence space in filter form (Definition 2.2) yields $\bigcap \mathcal{F}_x = N(x) \in \mathcal{F}_x$. □

Lemma 2.16. *Let $(X, \mathcal{T})$ be a topological space. Let $(X, \downarrow_{\mathcal{O}})$ be the induced convergence space in filter form with respect to the topology $\mathcal{T}$. We define $(\mathcal{F}, x) \in \mathcal{C}$ if and only if $\mathcal{F} \downarrow_{\mathcal{O}} x$. Then $\mathcal{C}$ is a convergence class in filter form.*

Proof. Since $(X, \downarrow_{\mathcal{O}})$ is a convergence space in filter form, conditions (a) and (b) in the definition of convergence class in filter form are already satisfied. We have to verify conditions (c) and (d). We have

$$(\mathcal{F}, x) \in \mathcal{C} \Longleftrightarrow \mathcal{F} \downarrow_{\mathcal{O}} x \Longleftrightarrow \mathcal{F} \supseteq N(x) \Longleftrightarrow \mathcal{F} \to x \text{ with respect to } \mathcal{T}.$$

First we prove condition (c). Let $(\mathcal{F}, s) \notin \mathcal{C}$. There exists $U \in N(s)$ with $U \notin \mathcal{F}$. Since $\mathcal{F}$ is a filter, for all $B \subseteq U$ we have $B \notin \mathcal{F}$ and we conclude that, for all $F \in \mathcal{F}$, we have $F \cap (X \setminus U) \neq \emptyset$. Let $\mathcal{B} = \{F \cap (X \setminus U) \,|\, F \in \mathcal{F}\}$. Then $\mathcal{B}$ is a base for a filter $\mathcal{F}' \supseteq \mathcal{F}$. Now let $\mathcal{G} \supseteq \mathcal{F}'$ be a filter on X. Assume that $U \in \mathcal{G}$. Because $\mathcal{B} \subseteq \mathcal{G}$, there exists $F \in \mathcal{F}$ with $F \cap (X \setminus U) \in \mathcal{B} \subseteq \mathcal{F}' \subseteq \mathcal{G}$ and $\emptyset = U \cap F \cap (X \setminus U) \in \mathcal{G}$, which is a contradiction. Thus, $U \notin \mathcal{G}$ and, for all filters $\mathcal{G} \supseteq \mathcal{F}'$, we get $(\mathcal{G}, s) \notin \mathcal{C}$, so that condition (c) is satisfied.

Condition (d) follows immediately because we defined $(\mathcal{F}, s) \in \mathcal{C}$ if and only if $\mathcal{F} \to s$ with respect to $\mathcal{T}$. We have only to apply Theorem 2.8 to complete the proof. □

Remark 2.2. Lemma 2.11 and Lemma 2.12 resp. Lemma 2.14 and Lemma 2.16 imply that, for every convergence class $\mathcal{C}$ for X in net form resp. filter form, there exists a topology $\mathcal{T}$ on X such that the induced convergence space is the convergence class $\mathcal{C}$, and that every topology $\mathcal{T}$ on X induces a topological convergence space in net form resp. filter form which is a convergence class in net form resp. filter form. Thus, the class of topological convergence spaces in net form resp. filter form is exactly the class of all convergence classes in net form resp. filter form. Indeed, the two conditions (c) and (d), in either definition of a convergence class, give the conditions needed for the convergence structure provided by a convergence space (in net form) (resp. in filter form) to be topological.

2.6. Interchange of Form in Convergence Classes

In the following, we will denote the closure operator of a convergence class in net form by ^{c_1}, and the closure operator of a convergence class in filter form by ^{c_2}. Let X be a non-empty set and let $\mathcal{C}$ be a convergence class on X in net form. By Theorem 2.7, ^{c_1} defines a topology $\mathcal{T}$ on X with the property $(S, s) \in \mathcal{C}$ if and only if $S_n \to s$ with respect to $\mathcal{T}$. As we have just seen in Lemma 2.16, $\mathcal{T}$ defines a convergence class $\mathcal{C}'$ on X in filter form with the property

$$(\mathcal{F}, s) \in \mathcal{C}' \iff \mathcal{F} \downarrow_{\mathcal{O}} s \iff \mathcal{F} \supseteq N(s) \iff \mathcal{F} \to s \text{ with respect to } \mathcal{T}.$$

Thus, the induced topologies of $\mathcal{C}$ and $\mathcal{C}'$ are equal, that is, we have $A^{c_1} = A^{c_2}$ for all $A \subseteq X$. In other words, we have for all $A \subseteq X$

$$\begin{aligned} A^{c_1} &= \{s \in X \mid \text{there is a net } S \text{ in } A \text{ s.t. } (S, s) \in \mathcal{C}\} \\ &= \{s \in X \mid \text{there is a filter } \mathcal{F} \text{ s.t. } A \in \mathcal{F} \text{ and } (\mathcal{F}, s) \in \mathcal{C}'\} = A^{c_2}. \end{aligned}$$

In fact, we can construct $\mathcal{C}'$ from $\mathcal{C}$ by means of the following result.

Lemma 2.17. *Let $X, \mathcal{C}$ and $\mathcal{C}'$ be as defined above. Then we have that $(\mathcal{F}, s) \in \mathcal{C}'$ if and only if there exists a net $S : D \to X$ such that $(S, s) \in \mathcal{C}$ and $\mathcal{B} = \{B_n \mid n \in D\}$ is a base for $\mathcal{F}$, where $B_n = \{S_m \mid m \geq n\}$.*

Proof. For sufficiency, let $(S, s) \in \mathcal{C}$ and suppose that $O \in N(s)$ is open. Then there exists $n \in D$ such that $S_m \in O$ for all $m \geq n$, and we have $B_n \subseteq O$. Since $\mathcal{F}$ is a filter, we get $O \in \mathcal{F}$ and therefore we have $N(s) \subseteq \mathcal{F}$. Thus, we conclude $(\mathcal{F}, s) \in \mathcal{C}'$ by the definition of $\mathcal{C}'$.

Conversely, let $(\mathcal{F}, s) \in \mathcal{C}'$. By definition, $\mathcal{F} \to s$ with respect to $\mathcal{T}$. Let $S : D_{\mathcal{F}} \to X$ be the net based on $\mathcal{F}$. By Theorem 2.1 (a), it follows that $S_n \to s$ with respect to $\mathcal{T}$. Thus, we obtain $(S, s) \in \mathcal{C}$ by means of the definition of $\mathcal{C}$. The construction of S yields $B_{(x,F)} = \{S_{(x',F')} \mid (x', F') \geq (x, F)\} = F$ for all $x \in F \in \mathcal{F}$. In particular, we have $\mathcal{B} = \{B_{(x,F)} \mid (x, F) \in D_{\mathcal{F}}\} = \mathcal{F}$ is a base for $\mathcal{F}$. □

Next, working in the opposite direction, we start with a convergence class $\mathcal{C}'$ for X in filter form. Let $\mathcal{T}$ be the induced topology, so that we have

$$(\mathcal{F}, s) \in \mathcal{C}' \iff \mathcal{F} \to s \text{ with respect to } \mathcal{T} \iff \mathcal{F} \supseteq N(s).$$

As we have seen in Lemma 2.12, $\mathcal{T}$ defines a convergence class $\mathcal{C}$ for X in net form and, by means of Lemma 2.10, the induced topologies of $\mathcal{C}$ and $\mathcal{C}'$ coincide and, hence, are equal to $\mathcal{T}$. Furthermore, $\mathcal{C}$ has the property

$(S, s) \in \mathcal{C}$ iff S is a net in X, $S : D \to X$ say, s.t. $S_n \to s$ relative to $\mathcal{T}$.

Once again Lemma 2.17 holds. In addition, we have the following result.

Lemma 2.18. *Let $X, \mathcal{C}, \mathcal{C}'$ and $\mathcal{T}$ be defined as above. Then we have $(S, s) \in \mathcal{C}$ iff there exists $(\mathcal{F}, s) \in \mathcal{C}'$ such that S is a subnet of T, where $D_{\mathcal{F}} = \{(x, F) \,|\, x \in F \in \mathcal{F}\}$ and T is the net $T : D_{\mathcal{F}} \to X$ defined by $T(x, F) = x$.*

Proof. For the necessity, let $(S, s) \in \mathcal{C}$, so that $S : D \to X$ is a net with $S_n \to s$ in the topology $\mathcal{T}$. Let $B_n = \{S_m \,|\, m \geq n\}$, and let $\mathcal{B} = \{B_n \,|\, n \in D\}$ be the base for the filter $\mathcal{F}$ generated by the net S. Using Theorem 2.1 (b), we see that $\mathcal{F} \to s$ with respect to $\mathcal{T}$. Thus, we conclude that $(\mathcal{F}, s) \in \mathcal{C}'$. Let $D_{\mathcal{F}}$ and T be as defined in our claim, and let $\varphi : D \to D_{\mathcal{F}}$ be defined by $\varphi(n) = (S_n, B_n)$ for all $n \in D$. Then, for all $n, m \in D$, we have that $n \leq m$ implies $B_m \subseteq B_n$ which in turn implies that $(S_n, B_n) \leq (S_m, B_m)$. So, φ is a monotonic mapping. Since, for all $x \in F \in \mathcal{F}$, there exists $n \in D$ such that $F \supseteq B_n$, we obtain $(x, F) \leq (S_n, B_n)$ so that the image of φ is cofinal in $D_{\mathcal{F}}$. Therefore, $T \circ \varphi$ is a net with $T(\varphi(n)) = T((S_n, B_n)) = S_n = S(n)$, that is, $S = T \circ \varphi$. Thus, S is a subnet of T, as required.

Conversely, let $(\mathcal{F}, s) \in \mathcal{C}'$. By definition of $\mathcal{C}'$, we have $\mathcal{F} \to s$ with respect to $\mathcal{T}$. Because T is the net based on $\mathcal{F}$, using Theorem 2.1 (a) we conclude that $T_n \to s$ with respect to $\mathcal{T}$. Since S is a subnet of T, we obtain also that $S_n \to s$ with respect to $\mathcal{T}$ so that $(S, s) \in \mathcal{C}$ by definition of $\mathcal{C}$. □

3. Convergence classes and VDM

As mentioned in the introduction, the paper falls naturally into two parts. The first of these is the previous section in which we established a rather satisfactory theory of convergence. This section constitutes the second part of the paper, and in it we want to apply certain of the convergence results of the first part to spaces of partial functions and to certain natural operators that they carry.

We begin by establishing some preliminaries and some notation.

3.1. Preliminaries

By the term *monoid*, we mean a non-empty set M endowed with a closed and associative binary operation $*$, called the law of composition or multiplication, which possesses an identity element u for the composition. There are several monoids of interest here, and two such examples of particular importance are $(\mathcal{P}(X), \cup, \emptyset)$ and $(\mathcal{P}(X), \cap, X)$, where $\mathcal{P}(X)$ again denotes the power set of X. In the first, the law of composition is the union of sets and the identity is the empty set; in the second the law of composition is the intersection of sets and the identity is the whole set X. We say that a monoid $(M, *, u)$ is a *topological monoid* if M is a topological space and the law of composition $*$ is a continuous function on $M \times M$, where $M \times M$ is endowed with the usual product topology determined by the topology on M.

We shall use the term *domain* (or *Scott domain*) with the meaning employed in [SLG94], which is our general reference to this subject. Thus, a *domain* $(D, \sqsubseteq, \perp)$, or simply D when no confusion is caused, is a consistently complete algebraic complete partial order. We let D_c denote the set of compact elements of D, and, given $x \in D$, we let $\text{approx}(x)$ denote the set $\{a \in D_c; a \sqsubseteq x\}$. Of course, $\text{approx}(x)$ is directed and $x = \sup \text{approx}(x)$ for each $x \in D$, where in general $\sup A$ denotes the supremum of the directed set A. Any complete partial order (cpo), and hence any domain, may be endowed with the well-known Scott topology, see [COMP80, SLG94], in which a set O is open if and only if it satisfies: (i) whenever $x \in O$ and $x \sqsubseteq y$, then $y \in O$, and (ii) whenever A is directed and $\sup A \in O$, then $A \cap O \neq \emptyset$. In the case of a domain, this topology has a rather simple description in that the collection $\{\uparrow a; a \in D_c\}$ is a basis for the Scott topology, where $\uparrow x = \{y \in D; x \sqsubseteq y\}$ for any $x \in D$.

One point of notation to which we should draw the attention of the reader is the following. In Section 2, we followed [Kel75] closely and therefore we used the symbol D for the index set of nets. From now on, since we are following [SLG94] closely, D will usually denote a domain, and therefore we will use I or J etc. to indicate index sets for nets and directed sets in general, and $i, j, n, m, \alpha, \beta$ etc. to denote elements of these index sets.

It will be useful to record next a couple of elementary facts we will use without further mention. The first is the well-known formulation of Scott continuity of functions between domains in terms of order properties, see [SLG94, Proposition 5.2.3]. *Let D and E be domains. Then a function $f : D \to E$ is continuous with respect to the Scott topology if and only if it satisfies the property: whenever $A \subseteq D$ is directed, we have that $f[A]$ is directed in E and $f(\sup A) = \sup f[A]$.* And the second concerns cartesian products, see [SLG94, Proposition 2.2.4]. *Let D, E and F be complete partial orders. Then a function $f : D \times E \to F$ is continuous if and only if f is continuous in each argument.*

Of course, the power set $\mathcal{P}(X)$ of a non-empty set X is a domain, ordered by set inclusion, whose compact elements are the finite sets. It is sometimes useful to identify $\mathcal{P}(X)$ with the set of all total functions from X to $\mathbf{2}$, by means of the characteristic functions of subsets of X, or with the product $\Pi_{i\in X}\,\mathbf{2}_i$ of X copies of $\mathbf{2}$, where $\mathbf{2}$ denotes the two-element set $\{0,1\}$. The usual product topology on $\Pi_{i\in X}\,\mathbf{2}_i$, when $\mathbf{2}$ is endowed with the Scott topology, results in the Scott topology on $\Pi_{i\in X}\,\mathbf{2}_i$ and hence on $\mathcal{P}(X)$. Alternatively, we may endow $\mathbf{2}$ with the discrete topology. Then, we will call the resulting topology on $\mathcal{P}(X)$, a *Cantor topology* since $\Pi_{i\in X}\,\mathbf{2}_i$ is homeomorphic to the Cantor set in the real line whenever X is denumerable. This topology also has significance in computing because of its well-known role in domain theory in relation to sets of maximal elements and universal domains. Moreover, it coincides with the Lawson topology on $\mathcal{P}(X)$ (the Lawson topology is the common refinement of the Scott topology and the lower topology, see [COMP80]). Finally, it has an important role in logic programming semantics (see [Sed95]) and in termination of logic programs, see [HS99]. In the present work, it turns out to be important in handling the override operator.

We have a simple characterization of net convergence in the Scott topology.

Proposition 3.1. *Let D be a domain. A net $x_i \to x$ in the Scott topology on D if and only if for each $a \in approx(x)$ there is an index i_0 such that $a \sqsubseteq x_i$ whenever $i_0 \le i$.*

Proof. Suppose that $x_i \to x$ and that $a \in \text{approx}(x)$. Then $\uparrow a$ is a Scott neighbourhood of x, and x_i eventually belongs to $\uparrow a$. So certainly there is an i_0 with the stated property.

Conversely, suppose that a net (x_i) and an element x are given in D and that the stated condition on elements $a \in \text{approx}(x)$ holds. Given an arbitrary Scott neighbourhood U of x, there is a basic neighbourhood $\uparrow a$ of x inside U, where $a \in \text{approx}(x)$. But the stated condition now simply asserts that there is an i_0 such that $x_i \in U$ whenever $i_0 \le i$ and therefore $x_i \to x$, as required. □

The following are simple, but useful, technical facts concerning the two topologies we have been discussing on $\mathcal{P}(X)$; the first follows from Proposition 3.1.

Proposition 3.2. *(1) In the Scott topology on $\mathcal{P}(X)$, a net A_i of sets converges to a set A if and only if every element of A is eventually an element of A_i.*
(2) In the Cantor topology on $\mathcal{P}(X)$, a net A_i converges to A if and only

if every element of A is eventually an element of A_i, and every element of X not in A is eventually not in A_i.

Example 3.3. *It is worth noting that, in each case, the stated conditions in Proposition 3.1 and in Proposition 3.2 actually define convergence classes generating the corresponding topology. Thus, they provide examples of topologies of interest in computing given in these terms.*

Proposition 3.4. *In either the Scott topology or the Cantor topology on $\mathcal{P}(X)$, both $(\mathcal{P}(X), \cup, \emptyset)$ and $(\mathcal{P}(X), \cap, X)$ are topological monoids.*

Proof. We show that $(\mathcal{P}(X), \cup, \emptyset)$ is a topological monoid in the Cantor topology, the arguments for the other claims being similar.

Suppose that $(A, B)_i = (A_i, B_i)$ is a net in $\mathcal{P}(X) \times \mathcal{P}(X)$ converging to (A, B) in the product of the Cantor topologies on $\mathcal{P}(X)$, thus A_i converges to A and B_i converges to B in $\mathcal{P}(X)$. Suppose $x \in A \cup B$. If $x \in A$, then, by Proposition 3.2, x is eventually in A_i and hence x is eventually in $A_i \cup B_i$, and similarly if it is the case that $x \in B$. If $x \notin A \cup B$, then x is not in A and x is not in B. Hence, by Proposition 3.2 again, x is eventually not in A_i and is eventually not in B_i, and hence x is eventually not in $A_i \cup B_i$. Thus, $A_i \cup B_i$ converges to $A \cup B$, and so $\cup$ is continuous, as required. □

The following result is proved similarly, and we omit the proof.

Proposition 3.5. *The mapping comp : $\mathcal{P}(X) \to \mathcal{P}(X)$ determined by taking the complement of a set and defined by comp$(S) = X \setminus S$ is continuous in the Cantor topology. Indeed, in the Cantor topology this mapping is a homeomorphism of $\mathcal{P}(X)$ onto itself and is an isomorphism between the topological monoids $(\mathcal{P}(X), \cup, \emptyset)$ and $(\mathcal{P}(X), \cap, X)$.*

Since comp is not even monotonic, it is clearly not Scott continuous on $(\mathcal{P}(X), \subseteq)$. It results from this that the override operator is not Scott continuous.

Definition 3.1. Let M be a topological monoid and let D be a domain which is also a topological space. Then M will be said to act on (the left of) D if there is a continuous function $\odot : M \times D \to D$, usually written $(m, x) \mapsto m \odot x$, with the following properties:
(i) $u \odot x = x$ for all $x \in D$.
(ii) $m_1 \odot (m_2 \odot x) = (m_1 * m_2) \odot x$ for all $m_1, m_2 \in M$ and all $x \in D$.
(iii) $m \odot a \in D_c$ for all $m \in M$ and all $a \in D_c$.

Given an action of M on D, fixing $m \in M$ determines a continuous function $x \mapsto m \odot x$ of D to itself which preserves the compact elements. Similarly, fixing $x \in D$ determines a continuous function $m \mapsto m \odot x$ from M to D.

3.2. The Basic Operators in VDM and in VDM$^\clubsuit$

Let X and Y be sets, and let $(X \to Y)$ denote the set of partial functions mapping X to Y. It is well-known that $(X \to Y)$ is a domain when ordered by graph inclusion: $\mu \sqsubseteq \nu$ if and only if $\mathrm{graph}(\mu) \subseteq \mathrm{graph}(\nu)$, where $\mathrm{graph}(\mu) = \{(x, y) \in X \times Y; x \in \mathrm{dom}(\mu) \text{ and } y = \mu(x)\}$, and here and elsewhere $\mathrm{dom}(\mu)$ denotes the domain of μ. Moreover, if $A = \{\mu_\alpha; \alpha \in I\}$ is a directed set of elements of $(X \to Y)$, then the supremum of A is the partial function well-defined by the union of the graphs of the $\mu_\alpha, \alpha \in I$. Finally, the compact elements of $(X \to Y)$ are the partial functions μ for which $\mathrm{graph}(\mu)$ is a finite set. We shall always suppose that $(X \to Y)$ is ordered in the way just described. However, we will need to endow $(X \to Y)$ with topologies other than the Scott topology, as well as with the Scott topology, in what follows. At any given time, unless stated otherwise, subsets of $(X \to Y) \times (X \to Y)$ will be given the subspace topology of the product with itself of whatever topology we are considering at that time on $(X \to Y)$.

For convenience, we state next a simple criterion for convergence in the Scott topology on $(X \to Y)$ which follows immediately from Proposition 3.1.

Proposition 3.6. *A net μ_i converges to μ in the Scott topology on $(X \to Y)$ if and only if whenever $(x, y) \in graph(\mu)$ we have $(x, y) \in graph(\mu_i)$ eventually.*

The operators which occur in VDM$^\clubsuit$ are operators defined on $(X \to Y)$. As already noted, it is our aim to study them from the domain-theoretic point of view and to determine the extent to which they are Scott continuous or otherwise. In fact, we work rather more generally than this since we formulate the results in terms of (continuous) actions of monoids on $(X \to Y)$, and obtain the results relative to the usual operators in VDM$^\clubsuit$ by fixing one or other of the arguments. It will be convenient to break the discussion into two parts, namely, into those which are Scott continuous, and those which are not. For general references to the details of the operator calculus used in VDM$^\clubsuit$, we cite [Mac90, Hug01]. In fact, the basic operators we study here are common to both VDM and VDM$^\clubsuit$.

3.3. Scott-Continuous Operators

In this subsection, the term "continuous" will mean Scott continuous unless otherwise stated.

The Extension Operator, $\sqcup$. Let μ and ν be elements of $(X \to Y)$ which satisfy $\mathrm{dom}(\mu) \cap \mathrm{dom}(\nu) = \emptyset$. We define the *extension* $\mu \sqcup \nu \in (X \to Y)$ *of* μ *by* ν as follows:

$$(\mu \sqcup \nu)(x) = \begin{cases} \mu(x) & \text{if } x \in \mathrm{dom}(\mu), \\ \nu(x) & \text{if } x \in \mathrm{dom}(\nu). \end{cases}$$

Theorem 3.7. *The mapping $(\mu, \nu) \mapsto \mu \sqcup \nu$ is Scott continuous as a mapping on the set $(X \to Y) \times_{\sqcup} (X \to Y)$ of those pairs (μ, ν) in $(X \to Y) \times (X \to Y)$ which satisfy $\mathrm{dom}(\mu) \cap \mathrm{dom}(\nu) = \emptyset$.*

Proof. We begin by showing that $(X \to Y) \times_{\sqcup} (X \to Y)$ is a subdomain of $(X \to Y) \times (X \to Y)$. Let $A = \{(\mu_\alpha, \nu_\alpha); \alpha \in I\}$ be a directed set in $(X \to Y) \times_{\sqcup} (X \to Y)$. Then A is directed as a subset of $(X \to Y) \times (X \to Y)$. Hence, $\pi_0[A]$ and $\pi_1[A]$ are directed sets in $(X \to Y)$, where π_0 and π_1 are the projections on the first and second factors respectively. Moreover, the supremum of A in $(X \to Y) \times (X \to Y)$ is the pair $(\sup \pi_0[A], \sup \pi_1[A])$, see [SLG94, Lemma 2.2.2]. We show that this pair belongs to $(X \to Y) \times_{\sqcup} (X \to Y)$ and hence that it is the supremum of A in $(X \to Y) \times_{\sqcup} (X \to Y)$. Suppose that $\mathrm{dom}(\sup \pi_0[A]) \cap \mathrm{dom}(\sup \pi_1[A]) \neq \emptyset$, and that x is an element of this non-empty set. Then there are α and β in I such that $(\sup \pi_0[A])(x) = \mu_\alpha(x)$ and $(\sup \pi_1[A])(x) = \nu_\beta(x)$. Since A is directed, there is $\gamma \in I$ with $(\mu_\alpha, \nu_\alpha) \sqsubseteq (\mu_\gamma, \nu_\gamma)$ and $(\mu_\beta, \nu_\beta) \sqsubseteq (\mu_\gamma, \nu_\gamma)$. But this leads to the conclusion that $(\sup \pi_0[A])(x) = \mu_\gamma(x)$ and $(\sup \pi_1[A])(x) = \nu_\gamma(x)$ and hence to the contradiction that $\mathrm{dom}(\mu_\gamma) \cap \mathrm{dom}(\nu_\gamma) \neq \emptyset$. Thus, it now follows that $(X \to Y) \times_{\sqcup} (X \to Y)$ is a subcpo of $(X \to Y) \times (X \to Y)$ and indeed it is readily checked that it is in fact a subdomain.

For the stated continuity, it suffices, by symmetry, to check continuity in either argument, so fix μ and consider the map $\nu \mapsto \mu \sqcup \nu$. If (μ, ν_1) and (μ, ν_2) are elements of $(X \to Y) \times_{\sqcup} (X \to Y)$ and $\nu_1 \sqsubseteq \nu_2$, then it is easily seen that $\mu \sqcup \nu_1 \sqsubseteq \mu \sqcup \nu_2$ so that the map in question is monotonic.

Now suppose that $A = \{\nu_\alpha; \alpha \in I\}$ is a directed set in $(X \to Y)$, where $\mathrm{dom}(\mu) \cap \mathrm{dom}(\nu_\alpha) = \emptyset$ for all $\alpha \in I$, and let $\nu = \sup A$. By the first part of the proof, we know that $\mathrm{dom}(\mu) \cap \mathrm{dom}(\nu) = \emptyset$. If $x \in \mathrm{dom}(\mu)$, then clearly $(\mu \sqcup \sup_{\alpha \in I} \nu_\alpha)(x) = (\mu \sqcup \nu)(x) = \mu(x)$. On the other hand, $(\mu \sqcup \nu_\alpha)(x) = \mu(x)$ for all α and hence $(\sup_{\alpha \in I}(\mu \sqcup \nu_\alpha))(x) = \mu(x)$. If $x \in \mathrm{dom}(\nu)$, then $\nu(x) = \nu_\beta(x)$ for some $\beta \in I$, and so $(\mu \sqcup \nu)(x) = \nu(x) = \nu_\beta(x) = (\mu \sqcup \nu_\beta)(x) = (\sup_{\alpha \in I}(\mu \sqcup \nu_\alpha))(x)$. Thus, $\mu \sqcup \sup_{\alpha \in I} \nu_\alpha = \sup_{\alpha \in I}(\mu \sqcup \nu_\alpha)$, and we have the required continuity. $\square$

The Glueing Operator, $\cup$. Let μ and ν be elements of $(X \to Y)$ which coincide on the intersection of their domains. Then μ may be *glued* to ν to obtain the partial map $\mu \cup \nu \in (X \to Y)$ defined as follows:

$$(\mu \cup \nu)(x) = \begin{cases} \mu(x) & \text{if } x \in \mathrm{dom}(\mu), \\ \nu(x) & \text{if } x \in \mathrm{dom}(\nu). \end{cases}$$

Theorem 3.8. *The mapping $(\mu, \nu) \mapsto \mu \cup \nu$ is Scott continuous as a mapping on the set $(X \to Y) \times_{\cup} (X \to Y)$ of those pairs (μ, ν) in $(X \to Y) \times (X \to Y)$ which coincide on the intersection of their domains.*

Proof. The proof of this result is similar to the proof of the previous result, and we omit the details. □

The Domain Restriction Operator, $\lhd$. Given $\mu \in (X \to Y)$ and an element S of $\mathcal{P}(X)$, we define the *restriction of μ by S* to be the partial function in $(X \to Y)$, denoted by $\lhd_S \mu$, which satisfies: (i) $\mathrm{dom}(\lhd_S \mu) = S \cap \mathrm{dom}(\mu)$, and (ii) $\lhd_S \mu$ coincides with μ on $S \cap \mathrm{dom}(\mu)$.

Theorem 3.9. *Suppose that $(X \to Y)$ is endowed with the Scott topology and that $\mathcal{P}(X)$ is endowed with either (1) the Scott topology, or (2) the Cantor topology. Then in either case, the mapping $\lhd : \mathcal{P}(X) \times (X \to Y) \to (X \to Y)$ defined by $\lhd(S, \mu) = \lhd_S \mu$ determines an action of the topological monoid $(\mathcal{P}(X), \cap, X)$ on the domain $(X \to Y)$.*

Proof. That $(\mathcal{P}(X), \cap, X)$ is a topological monoid in either case was shown in Proposition 3.4. If S_1 and S_2 are elements of $\mathcal{P}(X)$ and $\mu \in (X \to Y)$, then $S_1 \cap (S_2 \cap \mathrm{dom}(\mu)) = (S_1 \cap S_2) \cap \mathrm{dom}(\mu)$, and it follows that (ii) of Definition 3.1 is satisfied. The other two statements in this definition are clear, and so the result will follow as soon as we have established the required continuity.
(1) For this case, it suffices to check continuity in each argument separately.

Fix μ and consider the mapping $\theta : \mathcal{P}(X) \to (X \to Y)$ defined by $\theta(S) = \lhd_S \mu$. If $S_1 \subseteq S_2$, then $\mathrm{graph}(\lhd_{S_1} \mu) \subseteq \mathrm{graph}(\lhd_{S_2} \mu)$, and so θ is monotonic. Suppose that $A = \{S_\alpha; \alpha \in I\}$ is a directed family of sets in $\mathcal{P}(X)$ and let $S = \sup A = \bigcup_{\alpha \in I} S_\alpha$. We want to establish that $\lhd_{\bigcup_{\alpha \in I} S_\alpha} \mu = \sup_{\alpha \in I}(\lhd_{S_\alpha} \mu)$, for which it suffices to show that $\mathrm{graph}(\lhd_{\bigcup_{\alpha \in I} S_\alpha} \mu) = \bigcup_{\alpha \in I} \mathrm{graph}(\lhd_{S_\alpha} \mu)$ and this is a straightforward calculation.

Now fix S and consider the mapping $\phi : (X \to Y) \to (X \to Y)$ defined by $\phi(\mu) = \lhd_S \mu$. If $\mu_1 \sqsubseteq \mu_2$, that is, $\mathrm{graph}(\mu_1) \subseteq \mathrm{graph}(\mu_2)$, then it is clear that $\mathrm{graph}(\lhd_S \mu_1) \subseteq \mathrm{graph}(\lhd_S \mu_2)$ so that ϕ is monotonic. Suppose that $A = \{\mu_\alpha; \alpha \in I\}$ is a directed set in $(X \to Y)$, and let $\mu = \sup A$. We want to show that $\lhd_S \mu = \sup_{\alpha \in I}(\lhd_S \mu_\alpha)$, and again it is a straightforward calculation to show that $\mathrm{graph}(\lhd_S \mu) = \bigcup_{\alpha \in I} \mathrm{graph}(\lhd_S \mu_\alpha)$, which suffices.

(2) Suppose that the net $(S_i, \mu_i) \to (S, \mu)$ in the product space $\mathcal{P}(X) \times (X \to Y)$, where $\mathcal{P}(X)$ carries the Cantor topology. Then we have $S_i \to S$ in $\mathcal{P}(X)$ and $\mu_i \to \mu$ in $(X \to Y)$. We want to show that $\lhd_{S_i} \mu_i \to \lhd_S \mu$ in the Scott topology on $(X \to Y)$. Suppose that $\nu \in \text{approx}(\lhd_S \mu)$ is arbitrary. By Proposition 3.1, it suffices to show that there is an index i_0 such that $\nu \sqsubseteq \lhd_{S_i} \mu_i$ whenever $i_0 \leq i$ or, equivalently, that $\text{graph}(\nu) \subseteq \text{graph}(\lhd_{S_i} \mu_i)$ whenever $i_0 \leq i$. Since $\nu \in \text{approx}(\lhd_S \mu)$, we have $\text{graph}(\nu) \subseteq \text{graph}(\lhd_S \mu)$ and hence $\text{dom}(\nu) \subseteq \text{dom}(\lhd_S \mu)$ from which we obtain that $\text{dom}(\nu) \subseteq S$. But ν is a finite function, so that $\text{dom}(\nu)$ is a finite set, and $S_i \to S$ in the Cantor topology. Therefore, by applying Proposition 3.2 (2) as many times as there are elements in $\text{dom}(\nu)$, we see that there exists an index i_1 such that $\text{dom}(\nu) \subseteq S_i$ whenever $i_1 \leq i$.

Next, we note that $\nu \sqsubseteq \lhd_S \mu \sqsubseteq \mu$ and so $\nu \in \text{approx}(\mu)$. Since $\mu_i \to \mu$ in the Scott topology, by Proposition 3.1 there is an index i_2 such that $\nu \sqsubseteq \mu_i$ whenever $i_2 \leq i$. Choose an index i_0 such that $i_1 \leq i_0$ and $i_2 \leq i_0$. Then, whenever $i_0 \leq i$, we have $\text{graph}(\nu) \subseteq \text{graph}(\mu_i)$ and hence $\text{dom}(\nu) \subseteq \text{dom}(\mu_i)$. Since we also have $\text{dom}(\nu) \subseteq S_i$, we obtain $\text{dom}(\nu) \subseteq S_i \cap \text{dom}(\mu_i)$ and hence we obtain finally that $\text{graph}(\nu) \subseteq \text{graph}(\lhd_{S_i} \mu_i)$, or that $\nu \sqsubseteq \lhd_{S_i} \mu_i$, whenever $i_0 \leq i$, as required. $\square$

Notice that either part of this result implies that if we fix the set S, then the map $(X \to Y) \to (X \to Y)$ defined by $\mu \mapsto \lhd_S \mu$ is Scott continuous, and it is this map which is normally understood in the context of domain restriction within VDM♣.

3.4. Non-Scott-Continuous Operators

The Domain Removal Operator, ⩤. Given $\mu \in (X \to Y)$ and an element S of $\mathcal{P}(X)$, we define the *removal from* μ *of* S to be the partial function in $(X \to Y)$, denoted by $⩤_S \mu$, which satisfies: (i) $\text{dom}(⩤_S \mu) = \text{dom}(\mu) \setminus S$, and (ii) $⩤_S \mu$ coincides with μ on $\text{dom}(\mu) \setminus S$.

Thus, we now obtain a mapping $⩤ : \mathcal{P}(X) \times (X \to Y) \to (X \to Y)$ defined by $⩤(S, \mu) = ⩤_S \mu$. Moreover, if S_1 and S_2 are elements of $\mathcal{P}(X)$ and $\mu \in (X \to Y)$, then $(\text{dom}(\mu) \setminus S_2) \setminus S_1 = \text{dom}(\mu) \setminus (S_1 \cup S_2)$. Thus, (ii) of Definition 3.1 is satisfied, and (i) and (iii) also, relative to the monoid $(\mathcal{P}(X), \cup, \emptyset)$. Thus, algebraically, the mapping ⩤ determines an action of the monoid $(\mathcal{P}(X), \cup, \emptyset)$ on $(X \to Y)$. However, this mapping clearly cannot be continuous when the Scott topology is placed on $\mathcal{P}(X)$ and on $(X \to Y)$, since for fixed μ the map $\mathcal{P}(X) \to (X \to Y)$: $S \mapsto ⩤(S, \mu)$ is not even monotone, although it is clearly anti-monotone.

However, we note that for any $S \in \mathcal{P}(X)$ and any $\mu \in (X \to Y)$, we have the identity $⩤(S, \mu) = \lhd(X \setminus S, \mu) = \lhd(\text{comp}(S), \mu)$, so that

$\lessdot_S \mu = \lhd_{X \setminus S} \mu$. Since comp is an isomorphism of topological monoids by Proposition 3.5, it transforms the action of $\lessdot$ into an action of $\lhd$, and we immediately obtain from Theorem 3.9 the following result.

Theorem 3.10. *Suppose that $\mathcal{P}(X)$ is endowed with the Cantor topology and that $(X \to Y)$ is endowed with the Scott topology. Then the mapping $\lessdot : \mathcal{P}(X) \times (X \to Y) \to (X \to Y)$ defined by $\lessdot(S, \mu) = \lessdot_S \mu$ determines an action of the topological monoid $(\mathcal{P}(X), \cup, \emptyset)$ on the domain $(X \to Y)$.*

Thus, for any fixed S, the mapping $(X \to Y) \to (X \to Y) : \mu \mapsto \lessdot_S \mu$ is Scott continuous, and it is this map which is normally understood in the context of domain removal within VDM♣. Nevertheless, we have decided to discuss $\lessdot$ in this subsection, rather than in the previous one, because, for any fixed μ, the mapping $\mathcal{P}(X) \to (X \to Y)$: $S \mapsto \lessdot(S, \mu)$ is not continuous when $\mathcal{P}(X)$ and $(X \to Y)$ both carry the Scott topology, as already noted, although it is when $\mathcal{P}(X)$ carries the Cantor topology.

The Override Operator, †. Given $\mu, \nu \in (X \to Y)$, we define the partial map $\mu \dagger \nu \in (X \to Y)$, called the *override of* μ *by* ν, as follows:

$$(\mu \dagger \nu)(x) = \begin{cases} \nu(x) & \text{if } x \in \mathrm{dom}(\nu), \\ \mu(x) & \text{if } x \in \mathrm{dom}(\mu) \setminus \mathrm{dom}(\nu). \end{cases}$$

Thus, we obtain a mapping $\dagger : (X \to Y) \times (X \to Y) \to (X \to Y)$ defined by $\dagger(\mu, \nu) = \mu \dagger \nu$. Fixing the second argument ν, the mapping $\mu \mapsto \mu \dagger \nu$ can easily be seen to be Scott continuous by the methods we have been using thus far, and we omit the details. However, if we fix the first argument μ, and consider the mapping $\nu \mapsto \mu \dagger \nu$, it is easy to see that this mapping is not monotonic and hence is not Scott continuous. This has the consequence that $\dagger$ is not Scott continuous on $(X \to Y) \times (X \to Y)$. In fact, $\dagger$ does have certain continuity properties involving both the Cantor and Scott topologies which become apparent when one considers the canonical decomposition of $\mu \dagger \nu$ given below. Nevertheless, a satisfactory treatment of override appears to require a topology which suitably refines both the Scott and Cantor topologies, and we intend to introduce a satisfactory candidate for this shortly. However, before doing this we first establish the following result.

Proposition 3.11. *In the Scott topologies on $(X \to Y)$ and on $\mathcal{P}(X)$, the mapping dom: $(X \to Y) \to \mathcal{P}(X)$ defined by $\mu \mapsto$ dom(μ) is continuous.*

Proof. If $\mu_1 \sqsubseteq \mu_2$, that is, $\mathrm{graph}(\mu_1) \subseteq \mathrm{graph}(\mu_2)$, then $\mathrm{dom}(\mu_1) \subseteq \mathrm{dom}(\mu_2)$ and so dom is monotonic. Suppose that $A = \{\mu_\alpha; \alpha \in I\}$ is a directed set in $(X \to Y)$, and let $\mu = \sup_{\alpha \in I} \mu_\alpha$. We must show that $\mathrm{dom}(\mu) = \mathrm{dom}(\sup_{\alpha \in I} \mu_\alpha) = \bigcup_{\alpha \in I} \mathrm{dom}(\mu_\alpha)$. But μ is the partial function determined by the union of the graphs of the partial functions μ_α, and so the required equality is easily established, and dom is continuous. □

However, dom is not continuous if the Scott topology on $\mathcal{P}(X)$ is replaced by the Cantor topology. To see this, take Y to be the set of natural numbers, and take a net A_i which converges to A, say, in the Scott topology on $\mathcal{P}(X)$ but does not converge in the Cantor topology. Define $\mu_i \in (X \to Y)$ by setting $\mu_i(x)$ to be equal to 1 for all $x \in A_i$ and is otherwise undefined. Similarly, define $\mu \in (X \to Y)$ by setting $\mu(x)$ to be equal to 1 for all $x \in A$ and is otherwise undefined. Then $\mathrm{dom}(\mu_i) = A_i \not\to A = \mathrm{dom}(\mu)$ in the Cantor topology. However, it is a simple application of Proposition 3.1 and of the ideas used in the proof of (2) of Theorem 3.9 to see that $\mu_i \to \mu$ in the Scott topology on $(X \to Y)$ so that dom is not continuous. This fact is unfortunate and has significant bearing on subsequent developments.

Using the operators we have introduced so far, we can represent $\mu \dagger \nu$ by means of the equality $\mu \dagger \nu = \lhd_{\mathrm{dom}(\nu)} \mu \sqcup \nu$. This representation allows us to canonically decompose $\mu \dagger \nu$ into a composite of three mappings, in the following way.

(1) The first of the factors is the mapping $(X \to Y) \times (X \to Y) \to \mathcal{P}(X) \times (X \to Y) \times (X \to Y) : (\mu, \nu) \mapsto (\mathrm{dom}(\nu), \mu, \nu)$. Up to a reordering of the components, this mapping is the product $[\mathrm{dom}, \mathrm{Id}] \times \mathrm{Id}$, where Id denotes the identity map.

(2) The second factor is the mapping $\mathcal{P}(X) \times (X \to Y) \times (X \to Y) \to (X \to Y) \times (X \to Y) : (\mathrm{dom}(\nu), \mu, \nu) \mapsto (\lhd_{\mathrm{dom}(\nu)} \mu, \nu)$, and is the product $\lhd \times \mathrm{Id}$.

(3) The third factor is the mapping $(X \to Y) \times (X \to Y) \to (X \to Y) : (\lhd_{\mathrm{dom}(\nu)} \mu, \nu) \mapsto \lhd_{\mathrm{dom}(\nu)} \mu \sqcup \nu$ and is the mapping $\sqcup$.

Thus, in addition to the comments already made about the topology we wish to place on $(X \to Y)$, it is clear that if such a topology makes each of the mappings above continuous, then it will make the override operator continuous. If we give $(X \to Y)$ the Scott topology, then the first of these mappings is continuous provided $\mathcal{P}(X)$ carries the Scott topology by Proposition 3.11, but not if $\mathcal{P}(X)$ carries the Cantor topology. On the other hand, the second of the factors in the decomposition above is continuous if $\mathcal{P}(X)$ carries the Cantor topology by Theorem 3.10, but not if it carries the Scott topology. The way forward appears to be to provide $(X \to Y)$ with a suitable topology which makes both dom and $\lhd$ continuous when $\mathcal{P}(X)$ carries the Cantor topology.

One possible candidate for a suitable topology on $(X \to Y)$ which springs to mind is the Lawson topology, much studied in the context of complete and continuous lattices, see [COMP80]. As already noted, this topology is the common refinement of the Scott and lower topologies and therefore has some computational significance. Moreover, it is compact Hausdorff, at least in the case of continuous lattices, see [COMP80, III.1.10]. To define this topology, one takes as basic open sets the sets $U \backslash \uparrow F$, where U is Scott open, F is a finite set and, of course, $\uparrow F = \{y; f \sqsubseteq y \text{ for some } f \in F\}$. As a topology on $(X \to Y)$, the Lawson topology appears, unfortunately, not to be satisfactory for our purposes. For one thing, it turns out that dom is not continuous in this topology when $\mathcal{P}(X)$ has the Cantor topology, as we will see later. For another, related, reason it does not seem to be possible to formulate easily in these terms when a partial function μ is undefined at a point x, say. To say this, one would like to say $\mu \in U \backslash \uparrow \{[x \to y]; y \in Y\}$, where U is some Scott open set and $[x \to y]$ denotes the maplet, or partial function, whose value at x is y and is otherwise undefined. Unfortunately, the set $U \backslash \uparrow \{[x \to y]; y \in Y\}$ is not necessarily open in the Lawson topology if Y is infinite.

To solve this problem, we define a topology by means of convergence classes as described in Section 2. Thus, let $\mathcal{C}$ denote the set of all pairs (μ_i, μ), where μ_i is a net in $(X \to Y)$ and μ is an element of $(X \to Y)$, which satisfy the following condition: $(\mu_i, \mu) \in \mathcal{C}$ if and only if
(i) whenever $x \in \text{dom}(\mu)$, eventually $(x, \mu(x)) \in \text{graph}(\mu_i)$, and
(ii) whenever $x \notin \text{dom}(\mu)$, eventually $x \notin \text{dom}(\mu_i)$.

Thus, μ_i converges $(\mathcal{C})$ to μ or $\lim_i \mu_i \equiv \mu$ $(\mathcal{C})$ if and only if the conditions (i) and (ii) are satisfied.

We show next that the conditions above define a convergence class.

Theorem 3.12. *The condition μ_i converges $(\mathcal{C})$ to μ or $\lim_i \mu_i \equiv \mu$ $(\mathcal{C})$ if and only if:*
(i) whenever $x \in$ dom(μ), eventually $(x, \mu(x)) \in$ graph(μ_i), and
(ii) whenever $x \notin$ dom(μ), eventually $x \notin$ dom(μ_i)
determines a convergence class on $(X \to Y)$.

Proof. We must check that the four conditions of Definition 2.5 are satisfied.

(a) Suppose that μ_i is a constant net with value μ. Then it is clear that μ_i converges $(\mathcal{C})$ to μ.

(b) Suppose that $(\mu_i)_{i \in I}$ converges $(\mathcal{C})$ to μ and that $(\nu_j)_{j \in J}$ is a subnet of $(\mu_i)_{i \in I}$ determined by the function $\varphi : J \to I$. Thus, for each $j \in J$, we have $\nu_j = \mu_{\varphi(j)}$, and for each $i_0 \in I$ there is $j_0 \in J$ such that whenever $p \geq j_0$ in J we have $\varphi(p) \geq i_0$ in I. Suppose that $x \in \text{dom}(\mu)$.

Then there is $i_0 \in I$ such that $(x, \mu(x)) \in \text{graph}(\mu_i)$ whenever $i \geq i_0$. Choose the corresponding j_0 as above, and suppose that $p \geq j_0$. Then $\varphi(p) \geq i_0$. Thus, $\text{graph}(\nu_p) = \text{graph}(\mu_{\varphi(p)})$. But $\varphi(p) \geq i_0$, and therefore $(x, \mu(x)) \in \text{graph}(\mu_{\varphi(p)})$. Hence, $(x, \mu(x)) \in \text{graph}(\nu_p)$ and so eventually $(x, \mu(x)) \in \text{graph}(\nu_j)$.

Now suppose that $x \notin \text{dom}(\mu)$. Then there is $i_0 \in I$ such that whenever $i \geq i_0$, we have $x \notin \text{dom}(\mu_i)$. Choose j_0 as above in the properties defining a subnet, and suppose that $p \geq j_0$. Then $\varphi(p) \geq i_0$. Thus, $\text{dom}(\nu_p) = \text{dom}(\mu_{\varphi(p)})$ and hence $x \notin \text{dom}(\nu_p)$. Thus, eventually $x \notin \text{dom}(\nu_j)$, as required. Consequently, ν_j converges $(\mathcal{C})$ to μ.

(c) Suppose that $(\mu_i)_{i \in I}$ does not converge $(\mathcal{C})$ to μ. Then one or other of the defining conditions for convergence $(\mathcal{C})$ is violated. Suppose that the first is violated. Thus, there is an $x \in \text{dom}(\mu)$ such that $(x, \mu(x))$ is frequently not in $\text{graph}(\mu_i)$. In other words, there is a cofinal subset J of I such that $(x, \mu(x)) \notin \text{graph}(\mu_i)$ whenever $i \in J$. Clearly $(\mu_i)_{i \in J}$ determines a subnet of $(\mu_i)_{i \in I}$, in the usual way, no subnet of which converges $(\mathcal{C})$ to μ.

If the second defining condition of convergence $(\mathcal{C})$ is violated, we proceed similarly, and hence the third requirement, (c), is satisfied.

(d) Suppose that I is a directed set, let J_m be a directed set for each $m \in I$ and let F be the product directed set $I \times \prod_{m \in I} J_m$. Let $r : F \to I \times_I \bigcup_{m \in I} J_m$ be defined by $r(m, f) = (m, f(m))$. Now suppose that $\mu(m, n) \in (X \to Y)$ for all $m \in I, n \in J_m$, and suppose that $\lim_m \lim_n \mu(m, n) \equiv \nu$ $(\mathcal{C})$, where $\nu \in (X \to Y)$. We must show that $\mu \circ r$ converges $(\mathcal{C})$ to ν.

Let $x \in \text{dom}(\nu)$. We must find $(m, f) \in F$ such that if $(p, g) \geq (m, f)$ in F, then $(x, \nu(x)) \in \text{graph}(\mu \circ r(p, g))$. Choose $m \in I$ so that $(x, \nu(x)) \in \text{graph}(\lim_n \mu(p, n))$ for each $p \geq m$. Now, $(x, \nu(x)) \in \text{graph}(\lim_n \mu(p, n))$ means that $x \in \text{dom}(\lim_n \mu(p, n))$ and also that $(\lim_n \mu(p, n))(x) = \nu(x)$; moreover $\mu(p, n)$ converges $(\mathcal{C})$ to $\lim_n \mu(p, n)$ (for fixed p, and varying n). Therefore, for each $p \geq m$ we can choose $f(p) \in J_p$ such that $(x, \lim_n \mu(p, n)(x)) \in \text{graph}(\mu(p, n))$ whenever $n \geq f(p)$ in J_p. Thus, $(x, \nu(x)) \in \text{graph}(\mu(p, n))$ for all $n \geq f(p)$ in J_p. If p is a member of I which does not follow m in I (i.e. $m \not\leq p$), then let $f(p)$ be an arbitrary member of J_p, so that f is now everywhere defined. Now, if $(p, g) \geq (m, f)$ in F, then $p \geq m$ so that $(x, \nu(x)) \in \text{graph}(\lim_n \mu(p, n))$ and since $g(p) \geq f(p)$ we have $(\lim_n \mu(p, n))(x) = \nu(x)$. Hence, $(x, \nu(x)) \in \text{graph}(\mu(p, g(p)))$, or in other words, $(x, \nu(x)) \in \text{graph}(\mu \circ r(p, g))$. Thus, $(x, \nu(x))$ is eventually in $\text{graph}(\mu \circ r(p, g))$, as required.

Now suppose that $x \notin \text{dom}(\nu)$. Again, we must find $(m, f) \in F$ such that if $(p, g) \geq (m, f)$, then $x \notin \text{dom}(\mu \circ r(p, g))$. Choose $m \in I$ such that $x \notin \text{dom}(\lim_n \mu(p,n))$ for each $p \geq m$. Since $\mu(p,n)$ converges $(\mathcal{C})$ to $\lim_n \mu(p,n)$ (for fixed p, and varying n), for each $p \geq m$ we can choose $f(p) \in J_p$ such that $x \notin \text{dom}(\mu(p,n))$ whenever $n \geq f(p)$. If p is a member of I which does not follow m, then again let $f(p)$ be arbitrary in J_p. Now, if $(p,g) \geq (m,f)$, then $p \geq m$ and hence $x \notin \text{dom}(\lim_n \mu(p,n))$ and hence $x \notin \text{dom}(\mu(p, g(p))$ since $g(p) \geq f(p)$. So, $x \notin \text{dom}(\mu \circ r(p,g))$, as required.

Thus, $\mu \circ r$ converges $(\mathcal{C})$ to ν, and the proof is complete. □

This theorem results in a topology on $(X \to Y)$ which we will refer to as the *strong Cantor topology*, and the following remarks will explain why this name has been chosen.

3.13. **Remark**

(1) The conditions specifying convergence $(\mathcal{C})$ in Theorem 3.12 are easily seen to be equivalent to the following:
(a) whenever $(x,y) \in \text{graph}(\mu)$, eventually $(x,y) \in \text{graph}(\mu_i)$,
(b) whenever $x \notin \text{dom}(\mu)$, eventually $x \notin \text{dom}(\mu_i)$, and
(c) whenever $x \in \text{dom}(\mu)$ and $(x,y) \notin \text{graph}(\mu)$, then eventually $x \in \text{dom}(\mu_i)$ and $(x,y) \notin \text{graph}(\mu_i)$.

(2) Any partial function in $(X \to Y)$ may be identified with its graph and hence with a subset of $X \times Y$, so that we have a natural inclusion $(X \to Y) \subseteq \mathcal{P}(X \times Y)$ inside the power set of $X \times Y$. Thus, any convergence class respectively topology on $\mathcal{P}(X \times Y)$ induces a convergence class respectively topology on $(X \to Y)$. There are three such related to the present discussion:

(i) Convergence of a net A_i to A in $\mathcal{P}(X \times Y)$ specified by "whenever $(x,y) \in A$ we eventually have $(x,y) \in A_i$". By Example 3.3 and Proposition 3.1 this induces the Scott topology on $\mathcal{P}(X \times Y)$ and, as a subspace, induces the Scott topology on $(X \to Y)$, see Proposition 3.6.

(ii) Convergence of a net A_i to A in $\mathcal{P}(X \times Y)$ specified by "whenever $(x,y) \in A$ we eventually have $(x,y) \in A_i$, and whenever $(x,y) \notin A$ we eventually have $(x,y) \notin A_i$". By Example 3.3, we obtain the Cantor topology on $\mathcal{P}(X \times Y)$. We shall refer to the induced topology on $(X \to Y)$ in this case as the *Cantor topology* on $(X \to Y)$. Of course, as far as $(X \to Y)$ is concerned, a net μ_i converges to μ in the Cantor topology if and only if whenever $(x,y) \in \text{graph}(\mu)$ we eventually have $(x,y) \in \text{graph}(\mu_i)$, and whenever $(x,y) \notin \text{graph}(\mu)$ we eventually have $(x,y) \notin \text{graph}(\mu_i)$.

(iii) Thinking of a subset A of $X \times Y$ as a relation from X to Y, we define the *domain* of A, $\text{dom}(A)$, to be the set $\{x \in X; (x,y) \in A \text{ for some } y \in Y\}$. We can therefore generalize the conditions stated in (1), and hence those stated in Theorem 3.12, by specifying convergence of A_i to A in

$\mathcal{P}(X \times Y)$ to mean:
(a)$'$ whenever $(x, y) \in A$, eventually $(x, y) \in A_i$,
(b)$'$ whenever $x \notin \text{dom}(A)$, eventually $x \notin \text{dom}(A_i)$, and
(c)$'$ whenever $x \in \text{dom}(A)$ and $(x, y) \notin A$, then eventually $x \in \text{dom}(A_i)$ and $(x, y) \notin A_i$.
Following the proof of Theorem 3.12, it can be shown that this notion of convergence also defines a convergence class. Of course, the topology it induces on $(X \to Y)$ is the strong Cantor topology.
(3) It follows easily from the previous remark (1) that the convergence just defined, namely (2) (iii), implies that in (2) (ii) so that the strong Cantor topology is a refinement of the Cantor topology, and hence the name "strong Cantor topology".

We next collect together some basic facts about the topologies we have been discussing.

Proposition 3.14. *The following facts hold.*
(1) The strong Cantor topology is a refinement of the Cantor topology which in turn is a refinement of the Scott topology.
(2) The set $(X \to Y)$ is closed in $\mathcal{P}(X \times Y)$ in each of the three topologies on $\mathcal{P}(X \times Y)$ discussed in Remark 3.13.
(3) The space $(X \to Y)$ is compact and T_0 in the Scott topology.
(4) The space $(X \to Y)$ is compact Hausdorff in the Cantor topology.
(5) The space $(X \to Y)$ is Hausdorff in the strong Cantor topology and is compact if and only if the Cantor and strong Cantor topologies coincide.
(6) In general, the Cantor and strong Cantor topologies do not coincide, and therefore the strong Cantor topology is not generally compact.
(7) The strong Cantor topology is not trivial, that is, it is not the discrete topology.
(8) The space Y^X of all total functions mapping X into Y is not a closed subset of $(X \to Y)$ in the Scott and Cantor topologies, but is closed in the strong Cantor topology. In each of the three topologies in question, the induced topology on Y^X is not trivial, that is, is not discrete.
(9) The strong Cantor and Cantor topologies coincide on the set Y^X of all total functions in $(X \to Y)$.

Proof. (1) By Remark 3.13, the net convergence condition describing the strong Cantor topology is more restrictive than that describing the Cantor topology which in turn is more restrictive than that describing the Scott topology, and this observation suffices.
(2) Let A_i be a net of sets in $\mathcal{P}(X \times Y)$, each element of which is the graph of a partial function in $(X \to Y)$ and suppose that $A \in \mathcal{P}(X \times Y)$. By Remark 3.13, if A_i converges to A in any of the topologies in question, at least the following condition holds: whenever $(x, y) \in A$, eventually

$(x, y) \in A_i$. Thus, if (x, y') is also an element of A, then eventually it too is an element of A_i. But then it is immediate that $y = y'$ and hence that A is the graph of a partial function in $(X \to Y)$, as required.
(3) This is a well-known general fact about the Scott topology.
(4) The space $\mathcal{P}(X \times Y)$ is compact and Hausdorff in the Cantor topology. Hence, $(X \to Y)$ is also Hausdorff and, being closed by (2), is also compact.
(5) Since the strong Cantor topology is a refinement of the Cantor topology, it also is Hausdorff and so therefore is $(X \to Y)$ in this topology. If the Cantor and strong Cantor topologies agree, then obviously $(X \to Y)$ is also compact in the strong Cantor topology, by (2) and (4). On the other hand, if $(X \to Y)$ is compact in the strong Cantor topology, then the identity map regarded as a map from this space to $(X \to Y)$ with the Cantor topology is a one-to-one continuous mapping from a compact space onto a Hausdorff space, and therefore is a homeomorphism.
(6) Consider the partial functions $(\mathbb{N} \to \mathbb{N})$ from the set of natural numbers to itself. Define μ by setting $\mu(x) = 1$ if x is even, and taking μ to be undefined otherwise. For each $n \in \mathbb{N}$, define μ_n as follows: $\mu_n(x) = 1$ if x is even and $x \leq n$; $\mu_n(x) = n$ if x is odd and $x \leq n$; $\mu_n(x)$ is undefined if $x > n$. Then μ_n does not converge to μ in the strong Cantor topology because $\mathrm{dom}(\mu_n)$ does not converge to $\mathrm{dom}(\mu)$ in the Cantor topology on $\mathcal{P}(X)$, see Proposition 3.15 below. Yet μ_n does converge to μ in the Cantor topology. Thus, the Cantor and strong Cantor topologies are different in this case, and hence by the previous result (5) the space $(X \to Y)$ is not compact in the latter topology.
(7) Again consider $(\mathbb{N} \to \mathbb{N})$. Take μ as defined in (6). This time, for each natural number n, define μ_n by setting $\mu_n(x) = 1$ if x is even and $x \leq n$, and taking μ_n to be undefined otherwise. Then the sequence $(\mu_n)_n$ is not eventually constant and yet converges to μ in the strong Cantor topology.
(8) Take μ and μ_n as defined in (6) except that we now set $\mu_n(x) = 0$ if $x > n$, so that each of the μ_n is a total function. Again, μ_n does not converge to μ in the strong Cantor topology, but does so in the Scott and Cantor topologies. Since μ is not total, it follows that Y^X is not closed in $(X \to Y)$ in the Scott and Cantor topologies. On the other hand, if μ_n converges to μ in the strong Cantor topology and each of the μ_n is total, then so is μ by Proposition 3.15, and hence Y^X is closed in this case. Finally, take $(\mathbb{N} \to \mathbb{N})$ again, define μ by setting $\mu(x) = 1$ for all x and define μ_n by setting $\mu_n(x) = 1$ if $0 \leq x \leq n$ and setting $\mu_n(x) = 0$ otherwise; so that μ and each μ_n are total. Then $\mu_n \to \mu$ in the strong Cantor topology, and yet $(\mu_n)_n$ is not eventually constant, and so the induced topology on Y^X is not discrete.

(9) Let μ_i and μ be total functions in $(X \to Y)$. If $\mu_i \to \mu$ in the strong Cantor topology, then certainly $\mu_i \to \mu$ in the Cantor topology by (1). Conversely, suppose that $\mu_i \to \mu$ in the Cantor topology. Let $x \in \text{dom}(\mu)$. Then $(x, \mu(x)) \in \text{graph}(\mu)$ and so, by our current assumption, we have that $(x, \mu(x)) \in \text{graph}(\mu_i)$ eventually. The second convergence condition defining the strong Cantor topology is trivially satisfied in this case, and the result follows. □

Notice that (9) of Proposition 3.14 applies in particular when Y is the two-element set. Therefore, the strong Cantor and Cantor topologies coincide on the power set $\mathcal{P}(X)$. We will, however, persist in what follows in referring to the Cantor topology on $\mathcal{P}(X)$, rather than using the all-embracing term "strong Cantor topology".

We now proceed with our treatment of the override operator armed with the results that we have established at our disposal. We start with the following simple, but important, fact.

Proposition 3.15. *If μ_i converges to μ in the strong Cantor topology, then dom(μ_i) converges to dom(μ) in the Cantor topology on $\mathcal{P}(X)$. Hence, the map dom is continuous when $(X \to Y)$ is endowed with the strong Cantor topology and $\mathcal{P}(X)$ is endowed with the Cantor topology.*

Proof. Suppose that $\mu_i \to \mu$ in the strong Cantor topology. Let $x \in \text{dom}(\mu)$. Then eventually $(x, \mu(x)) \in \text{graph}(\mu_i)$ and hence eventually $x \in \text{dom}(\mu_i)$. On the other hand, if $x \notin \text{dom}(\mu)$, then eventually $x \notin \text{dom}(\mu_i)$. Thus, $\text{dom}(\mu_i) \to \text{dom}(\mu)$ in the Cantor topology on $\mathcal{P}(X)$, as required. □

3.16. **Remark**

(1) The examples in (6) and (8) of Proposition 3.14 show that the function dom is not continuous when $(X \to Y)$ and $\mathcal{P}(X)$ both have the Cantor topologies, so that the Cantor topology on $(X \to Y)$ is not entirely satisfactory. Since the Lawson topology on $\mathcal{P}(X \times Y)$ coincides with the Cantor topology, the Lawson topology is also not entirely satisfactory in this context.

(2) These comments raise the general question "Just what is a reasonable notion of convergence in $(X \to Y)$?", the answer depending of course on the ultimate applications. In subjects like functional analysis, for example, convergence is often uniform ($\sup_x ||f(x) - f_n(x)|| \to 0$ as $n \to \infty$ in some norm $||\cdot||$) or uniform convergence on compacta etc. These notions are intuitively very reasonable and do indeed capture the notion of the function f_n tending towards the function f. However, the price one pays for this is relatively few convergent sequences (or nets), hence relatively many open sets, hence the underlying topology is highly non-compact,

usually. By comparison, convergence in $(X \to Y)$ can be quite bizarre as shown by some of the examples given in the proof of Proposition 3.14. In fact, convergence of $\text{dom}(\mu_i)$ to $\text{dom}(\mu)$ as part of one's definition is quite natural and, intuitively, convergence in the strong Cantor topology is probably the most reasonable. However, again the price to be paid is non-compactness in general.

Notwithstanding the remarks just made about non-compactness, it will become apparent later that the following result shows that the strong Cantor topology is in many ways the best possible choice of topology to impose on $(X \to Y)$.

Proposition 3.17. *The strong Cantor topology is the smallest topology on $(X \to Y)$ which refines both the Scott topology and the Lawson topology and in which the function dom is continuous when $\mathcal{P}(X)$ is endowed with the Cantor (or Lawson) topology.*

Proof. By Proposition 3.14, the strong Cantor topology certainly refines both the Scott topology and the Lawson topology, and by Proposition 3.15 satisfies the continuity requirement concerning dom.

Now suppose $\mathcal{T}$ is any topology refining both the topologies mentioned and such that dom is continuous. Let $\mu_i \to \mu$ in $\mathcal{T}$. Then $\mu_i \to \mu$ in the Scott topology. Let $x \in \text{dom}(\mu)$. Then $(x, \mu(x)) \in \text{graph}(\mu)$. Hence, by Proposition 3.6, $(x, \mu(x)) \in \text{graph}(\mu_i)$ eventually. On the other hand, $\text{dom}(\mu_i) \to \text{dom}(\mu)$ in the Cantor topology on $\mathcal{P}(X)$. Hence, if $x \notin \text{dom}(\mu)$, then eventually $x \notin \text{dom}(\mu_i)$. Thus, $\mu_i \to \mu$ in the strong Cantor topology, as required. □

Theorem 3.18. *The mapping $\lhd : \mathcal{P}(X) \times (X \to Y) \to (X \to Y)$ defined by $\lhd(S, \mu) = \lhd_S \mu$ is continuous when $\mathcal{P}(X)$ is endowed with the Cantor topology and $(X \to Y)$ is endowed with either:*
(i) the strong Cantor topology, or
(ii) the Cantor topology.

Proof. We content ourselves by proving the result referring to (i) and leave the details of (ii), which are similar, to the reader.

Thus, suppose that $(S_i, \mu_i) \to (S, \mu)$ in $\mathcal{P}(X) \times (X \to Y)$. Thus, $S_i \to S$ in the Cantor topology on $\mathcal{P}(X)$ and $\mu_i \to \mu$ in the strong Cantor topology on $(X \to Y)$. We must show that $\lhd_{S_i} \mu_i \to \lhd_S \mu$ in the strong Cantor topology.

Suppose that $x \in \text{dom}(\lhd_S \mu)$. Thus, $x \in S$ and $x \in \text{dom}(\mu)$. Since $S_i \to S$, we eventually have $x \in S_i$; since $\mu_i \to \mu$, we eventually have $(x, \mu(x)) \in \text{graph}(\mu_i)$. Using the directedness of the index set of the net in question, we can choose the index i_0 to get these statements holding simultaneously beyond i_0. Thus, beyond i_0 we have $x \in S_i \cap \text{dom}(\mu_i)$

and $(x, \mu(x)) \in \mathrm{graph}(\mu_i)$. Hence, beyond i_0 we have $\mu_i(x) = \mu(x)$ and hence we have $\lhd_{S_i} \mu_i(x) = \lhd_S \mu(x)$. Thus, $(x, \lhd_S \mu(x)) \in \mathrm{graph}(\lhd_{S_i} \mu_i)$ beyond i_0 and, hence, eventually.

Now suppose that $x \notin \mathrm{dom}(\lhd_S \mu) = S \cap \mathrm{dom}(\mu)$.
Case 1: $x \notin S$. Since $S_i \to S$, eventually $x \notin S_i$ and hence eventually $x \notin S_i \cap \mathrm{dom}(\mu_i)$ so that eventually $x \notin \mathrm{dom}(\lhd_{S_i} \mu_i)$.
Case 2: $x \notin \mathrm{dom}(\mu)$. Since $\mu_i \to \mu$, eventually $x \notin \mathrm{dom}(\mu_i)$, and hence eventually $x \notin S_i \cap \mathrm{dom}(\mu_i)$, that is, eventually $x \notin \mathrm{dom}(\lhd_{S_i} \mu_i)$.

This covers all cases, and so now we see that $\lhd_{S_i} \mu_i \to \lhd_S \mu$, as required. □

Bearing in mind our earlier comments about comp transforming the action of $\unlhd$ into one of $\lhd$ and vice versa, we immediately obtain from Theorem 3.18 the following result.

Theorem 3.19. *The mapping* $\unlhd : \mathcal{P}(X) \times (X \to Y) \to (X \to Y)$ *defined by* $\unlhd(S, \mu) = \unlhd_S \mu$ *is continuous when* $\mathcal{P}(X)$ *is endowed with the Cantor topology and* $(X \to Y)$ *is endowed with either:*
(i) the strong Cantor topology, or
(ii) the Cantor topology.

Theorem 3.20. *The mapping* $\sqcup : (X \to Y) \times_{\sqcup} (X \to Y) \to (X \to Y)$ *defined by* $\sqcup(\mu, \nu) = \mu \sqcup \nu$ *is continuous when* $(X \to Y)$ *is endowed with either:*
(i) the strong Cantor topology, or
(ii) the Cantor topology.

Proof. Again, we prove just the first of these claims and leave the second to the reader.

Let (μ_i, ν_i) be a net converging in $(X \to Y) \times_{\sqcup} (X \to Y)$ to (μ, ν) relative to the strong Cantor topology. We must show that $\mu_i \sqcup \nu_i$ converges in this topology to $\mu \sqcup \nu$. Now $\mu_i \to \mu$ and $\nu_i \to \nu$. Suppose that $x \in \mathrm{dom}(\mu \sqcup \nu) = \mathrm{dom}(\mu) \cup \mathrm{dom}(\nu)$ and for sake of argument suppose that $x \in \mathrm{dom}(\mu)$. Then eventually $(x, \mu(x)) \in \mathrm{graph}(\mu_i)$ and hence eventually $(x, \mu(x)) \in \mathrm{graph}(\mu_i \sqcup \nu_i)$, since $(\mu_i \sqcup \nu_i)(x) = \mu_i(x) = \mu(x)$. Similarly, if $x \in \mathrm{dom}(\nu)$.

Now suppose that $x \notin \mathrm{dom}(\mu \sqcup \nu)$. Then $x \notin \mathrm{dom}(\mu)$ and $x \notin \mathrm{dom}(\nu)$. Thus, eventually we simultaneously get $x \notin \mathrm{dom}(\mu_i)$ and $x \notin \mathrm{dom}(\nu_i)$. Hence, eventually we get $x \notin \mathrm{dom}(\mu_i \sqcup \nu_i)$, as required. □

Recalling the canonical decomposition of the override operator and using the results just established we now obtain the following main result.

Theorem 3.21. *The mapping* $\dagger : (X \to Y) \times (X \to Y) \to (X \to Y)$ *defined by* $\dagger(\mu, \nu) = \mu \dagger \nu$ *is continuous in the strong Cantor topology.*

4. Compactness of $(X \to Y)$

To close, we consider the compactness of the space $(X \to Y)$ in the strong Cantor topology. To do this, let $\bot$ be some object not in Y, let $Y_\bot$ denote the set $Y \cup \{\bot\}$, and let $\bot^X$ denote the constant map on X with value $\bot$.

Proposition 4.1. *For any set* Y, *the Cantor topology on* Y^X *coincides with the topology of pointwise convergence, where* Y *has the discrete topology.*

Proof. Suppose that $f_i \to f$ in the Cantor topology on Y^X, and that $f(x) = y$. Then $(x, y) \in \text{graph}(f)$ and hence eventually we have $(x, y) \in \text{graph}(f_i)$. Thus, eventually $f_i(x) = y$ and so $f_i(x)$ converges to y in the discrete topology on Y. Conversely, suppose that $f_i \to f$ in the topology of pointwise convergence, where Y has the discrete topology. If $(x, y) \in \text{graph}(f)$, then $f(x) = y$ and so eventually $f_i(x) = y$, that is, eventually $(x, y) \in \text{graph}(f_i)$. Now suppose that $(x, y) \notin \text{graph}(f)$. Then $f(x) \neq y$, and so eventually $f_i(x) \neq y$ and consequently we have eventually that $(x, y) \notin \text{graph}(f_i)$. Hence, $f_i \to f$ in the Cantor topology on Y^X. □

Let $t : (X \to Y) \to Y_\bot$ be the mapping defined by $t(\mu) = \bot^X \dagger \mu$, and let p be the mapping $p : Y_\bot \to (X \to Y)$, where $p(f) = f'$ is the partial map f' obtained by the removal from the total map f of the set S of all those $x \in X$ such that $f(x) = \bot$. These mappings are bijections and each is the inverse of the other. Indeed, in the strong Cantor topology each is a homeomorphism as we see in the proof of the following result.

Proposition 4.2. *The space* $(X \to Y)$ *is compact in the strong Cantor topology if and only if* Y *is finite.*

Proof. When endowed with the strong Cantor topology, both $(Y_\bot)^X$ and $(X \to Y)$ are subspaces of $(X \to Y_\bot)$ in the strong Cantor topology. Therefore, on fixing the first factor of $\dagger$ at $\bot^X$ in Theorem 3.21, we see that t is continuous in the strong Cantor topology.

The map p is also continuous in the strong Cantor topology, as we now show. Suppose that $f_i \to f$ in the strong Cantor topology on $(Y_\bot)^X$. Let $x \in \text{dom}(f')$, so that $f'(x) = f(x) \neq \bot$. Since $f_i \to f$, we have $(x, f(x))$ eventually in $\text{graph}(f_i)$. Thus, eventually $f_i(x) = f(x) \neq \bot$. So, eventually $f_i'(x) = f'(x) \neq \bot$, that is, eventually $(x, f'(x)) \in \text{graph}(f_i')$. Now suppose that $x \notin \text{dom}(f')$. This means that $f(x) =$

$\perp$. Since $f_i \to f$, eventually $(x, f(x)) \in \mathrm{graph}(f_i)$ so that eventually $f_i(x) = f(x) = \perp$ and hence eventually $x \notin \mathrm{dom}(f'_i)$. Thus, $f'_i \to f'$ in the strong Cantor topology on $(X \to Y)$, as required. Thus, t and p are homeomorphisms.

Suppose now that $(X \to Y)$ is compact in the strong Cantor topology. Then $(Y_\perp)^X$ is compact in this topology and hence is compact in the Cantor topology by (9) of Proposition 3.14. But then $Y_\perp$ is finite by Proposition 4.1 and Tychonoff's theorem, and therefore Y is finite. Conversely, if Y is finite, then $(Y_\perp)^X$ is compact in the Cantor topology and hence in the strong Cantor topology, and therefore $(X \to Y)$ is compact in the strong Cantor topology. □

One can avoid the introduction of the element $\perp$, and we briefly consider the following alternative way of proceeding to obtain the previous result. To do this we must, however, choose a base point in Y. Having made this choice, u say, we then let u^X denote the constant function on X with value u.

Let $p : \mathcal{P}(X) \times Y^X \to (X \to Y)$ be defined by $p(S, f) = \lhd_S f$, noting that no confusion will be caused by use of the symbol p again. It follows immediately from Theorem 3.18 that p is continuous when Y^X and $(X \to Y)$ have the strong Cantor topology, and $\mathcal{P}(X)$ has the Cantor topology. Furthermore, there is a naturally defined section $s : (X \to Y) \to \mathcal{P}(X) \times Y^X$ of this mapping p, where $s(\mu) = (\mathrm{dom}(\mu), u^X \dagger \mu)$, and s is clearly continuous by virtue of Proposition 3.15 and the proof of the continuity of t given above.

The mappings p and s determine an endomorphism $e = s \circ p$ of the product space $\mathcal{P}(X) \times Y^X$ to itself which is continuous in the topologies under discussion. Since s is a section of p, it follows that e is idempotent and so $e(S, f)$ is a fixed point of e for each pair (S, f). Letting $F = \mathrm{fix}(e)$ denote the set of all fixed points of e, we note that F is a closed set in $\mathcal{P}(X) \times Y^X$, as is readily checked. We further note the following facts which are established in [Hug01], see also [LS97]: (i) $(S, f) \in F$ if and only if $f(x) = u$ for all $x \in X \setminus S$, and (ii) the restriction $p|_F$ of p to F is an inverse of s.

We can now give an alternative proof of Proposition 4.2, as follows. First, we note that $(X \to Y)$ and F are homeomorphic (under $p|_F$ or s) and F is closed. Therefore, if Y is finite, it is immediate from Proposition 4.1 that F is compact and hence that $(X \to Y)$ is compact. Conversely, if $(X \to Y)$ is compact, then F is compact. But $(X, f) \in F$ for each $f \in Y^X$. Therefore, the image of F under the projection on the second factor on the product $\mathcal{P}(X) \times Y^X$ is all of Y^X and hence the latter is compact. It follows that Y is finite.

The identification above of $(X \to Y)$ and $(Y_\perp)^X$ using t and p is the one customarily used except that we have expressed it in the language of VDM. However, t is not continuous in the Scott topology nor in the Cantor topology, and this is one of the reasons why the strong Cantor topology seems to be quite appropriate for carrying out functional calculus on spaces of partial functions. In addition it gives a smooth treatment of such spaces and their operators, as shown by the mechanics of the proofs above.

5. Conclusions and further work

The results of the first part of the paper give a very satisfactory development of convergence spaces and convergence classes in terms of both nets and filters. Moreover, we establish complete duality between the two theories, and present simple conditions under which a convergence space is topological. Specifically, a convergence space $(X,\mathcal{S})$ in net form is topological if and only if it is a convergence class in net form if and only if it satisfies conditions (c) and (d) of Definition 2.5. Similarly, a convergence space $(X,\mathcal{F})$ in filter form is topological if and only if it is a convergence class in filter form if and only if it satisfies conditions (c) and (d) of Definition 2.6.

The second part shows that in relation to VDM, the Scott topology is not satisfactory: certain of the standard basic operators encountered there are Scott continuous, others are not. Overcoming this has necessitated the introduction of (smallest possible) refinements of the Scott topology such as the strong Cantor topology and then all the basic operators considered are continuous, giving a satisfactory analysis.

Noting that "constructive" and "effective" are closely related concepts, and that "effective" and "continuous" are also closely related from the domain-theoretic point of view, it is of interest to investigate the effectiveness of the operators we have discussed within the topological framework of this paper. This objective is closely related to the programme being carried out in [Hug01, Mac90, Mac01], where the operators we have considered (particularly the override) have been studied in [Hug01, Mac90, Mac01] from the point of view of topos theory in order to view them constructively. Thus, one may view the present paper as taking a first step towards examining the effectiveness of the operators concerned by considering the possibility of constructive topology within VDM$^\clubsuit$ in the spirit of [Mac01]. Indeed, one of our objectives here has been to provide a "convenient category" in which all the operations considered are automatically continuous, and the strong Cantor topology essentially does this. However, the full objective of contributing an appropriate notion of effectiveness within the framework of [Mac01] is ongoing work of the authors and will be discussed elsewhere.

References

[BR01] H.A. Blair and J. Remmel, *Hybrid Automata: Convergence Spaces and Continuity.* Proceedings of the joint IIIS & IEEE meeting of the 5th World Multiconference on Systemics, Cybernetics and Informatics (SCI2001) and the 7th International Conference on Information Systems Analysis and Synthesis (ISAS2001), Orlando, Florida, USA, July, 2001. International Institute of Informatics and Systemics, 2001.

[COMP80] G. Gierz et al., *A Compendium of Continuous Lattices.* Springer-Verlag, Berlin, Heidelberg, New York, 1980.

[Hec00] R. Heckmann, *A Non-Topological View of Dcpo's as Convergence Spaces.* In: T. Hurley, M. Mac an Airchinnigh, M. Schellekens and A.K. Seda (Eds.), Proceedings of the First Irish Conference on the Mathematical Foundations of Computer Science and Information Technology, Cork, Ireland, July , 2000. Electronic Notes in Theoretical Computer Science, Volume 40, 2001, Elsevier, pp. 1–22.

[HS99] P. Hitzler and A.K. Seda, *Acceptable Programs Revisited.* Proceedings of the Workshop on Verification in Logic Programming, Sixteenth International Conference on Logic Programming (ICLP'99), Las Cruces, New Mexico, November, 1999. Electronic Notes in Theoretical Computer Science, Volume 30, No. 1, Elsevier, 1999, pp. 1–18.

[Hyl79] J.M.E. Hyland, *Filter Spaces and Continuous Functionals*, Annals of Mathematical Logic, 16 (1979), 101–143.

[Hug01] A. Hughes, *Elements of an Operator Calculus.* Ph.D thesis, Department of Computer Science, University of Dublin, Trinity College, Dublin, Ireland, 2001.

[Kel75] J.L. Kelley, *General Topology.* Graduate Texts in Mathematics, Vol. 27, Springer-Verlag, Berlin, 1975.

[LS97] F.W. Lawvere and S.H. Schanuel, *Conceptual Mathematics: A First Introduction to Categories.* Cambridge University Press, Cambridge, 1997.

[Mac90] M. Mac an Airchinnigh, *Conceptual Models and Computing.* Ph.D thesis, Department of Computer Science, University of Dublin, Trinity College, Dublin, Ireland, 1990.

[Mac01] M. Mac an Airchinnigh, *Towards a Topos Theoretic Foundation for the Irish School of Constructive Mathematics ($M_C^\clubsuit$).* In: J.N.

Oliveira and P. Zave (Eds.), Proceedings of FME 2001: Formal Methods for Increasing Software Productivity, Lecture Notes in Computer Science, Vol. 2021, Springer-Verlag, 2001.

[Pau87] L. Paulson, *Logic and Computation: Interactive Proof with Cambridge LCF.* Cambridge Tracts in Theoretical Computer Science Number 2, Cambridge University Press, 1987.

[Sed95] A.K. Seda, *Topology and the Semantics of Logic Programs*, Fundamenta Informaticae **24** (4) (1995), 359–386.

[Sed02] A.K. Seda, *Some Convergence Issues in Theoretical Computer Science.* Proceedings of Information'2002, Beijing, July, 2002, pp. 1–18. Information **5** (4) (2002), pp. 447–462.

[Smy92] M.B. Smyth, *Topology.* In: S. Abramsky, D.M. Gabbay and T.S.E. Maibaum (Eds.): Handbook of Logic in Computer Science, vol. 1, pp. 641–761. Clarendon Press, Oxford University, 1992.

[SLG94] V. Stoltenberg-Hansen, I. Lindström and R. Griffor, *Mathematical Theory of Domains.* Cambridge Tracts in Theoretical Computer Science Number 22, Cambridge University Press, 1994.

[Wil70] S. Willard, *General Topology.* Addison-Wesley, Reading MA, 1970.

Chapter 5

ON MEET-CONTINUOUS DCPOS*

Hui Kou, Ying-Ming Liu and Mao-Kang Luo
Department of Mathematics
Sichuan University
Chengdu 610064, China
kouhui@vip.sina.com, makaluo@scu.edu.cn

Abstract

It is well-known that a complete lattice L is a meet-continuous lattice if and only if $x \wedge \bigvee D = \bigvee_{d \in D} x \wedge d$ for all $x \in P$. This property in fact can be characterized by the Scott topology simply as $\mathrm{cl}_\sigma(\downarrow x \cap \downarrow D) = \downarrow x$ whenever $x \leq \bigvee D$. Since the meet operator is not involved, the topological property of meet-continuity can be naturally extended to general dcpos. Such dcpos are also called meet-continuous in this note. It turns out that there exist close relations among meet-continuity, Hausdorff separation, quasicontinuity, continuity and Scott-open filter bases. In particular, we prove that Hausdorff dcpos (via the Lawson topology) need not be quasicontinuous, the category **CONT** is not a reflective full subcategory of **QCONT**, the category of quasicontinuous domains, and a dcpo P is meet-continuous when $\sigma(P) = \bar{\sigma}(P)$ or it is a semilattice with a T_0 $\bar{\sigma}(D)$-topology, where $\bar{\sigma}(P)$ denote the topology generated by all the Scott-open filters of P. Moreover, under appropriate conditions, the category of dcpos with T_0 $\bar{\sigma}(D)$-topology form a cartesian closed category.

Keywords: Meet-continuous dcpo, quasi-continuous domain, continuous domain, Scott-open filter

*This work is supported by the Natural Science Foundation, the Mathematics Tianyuan Foundation and 973 Program of China.

GQ Zhang et al. (eds.), Domain Theory, Logic and Computation, 117-135.

1. Introduction

The theory of domains was established in order to have appropriate spaces on which to define semantic functions for the denotational approach to programming-language semantics. A large number of people have made essential contributions in this area and they have shown in particular that domain theory is not only theoretically important (see, for example, Scott [22] Gierz et al [3], Abramsky and Jung [1], Mislove [21]) but also practically useful (see, for example, Amadio and Curien [2], Gunter [5], Winskel [24]). There have been many profound results that characterize complicated domains and operators on them, such as FS-domains, bifinite domains, Scott domains, function spaces and powerdomains, and so on. But, even with respect to the most basic structures, i.e., dcpos, the story, however, is not finished. It is well known that a complete lattice L is a meet-continuous lattice if and only if $x \wedge \bigvee D = \bigvee_{d \in D} x \wedge d$ for all $x \in L$ and a directed set $D \subseteq L$. This property in fact can be characterized by the Scott topology simply as $\mathrm{cl}_\sigma(\downarrow x \cap \downarrow D) = \downarrow x$ whenever $x \leq \bigvee D$. Since the meet operator is not involved, the topological property of meet-continuity can be naturally extended to general dcpos. Such dcpos are also called meet-continuous in this note. It turns out that a dcpo D is meet-continuous if and only if the lattice $\Gamma(D)$ of Scott closed sets is a complete Heyting algebra. This means that a continuous dcpo is meet-continuous and a quasicontinuos dcpo is continuous if and only if it is a meet-continuous dcpo. For the Hausdorff separation, we show that if D, E are continuous dcpos with least elements, then $[D \longrightarrow E]$ is Hausdorff in its Lawson topology. So, if a continuous semilattice D with bottom is neither Lawson compact nor an L-domain, then the function space $[D \longrightarrow D]$ exactly provides an example to show that a dcpo with the Hausdorff Lawson topology need not be quasicontinuous, because $[D \longrightarrow D]$ is meet-continuous. This is a clear difference from the case of complete lattices. Note that since there are a number of quasicontinuous dcpos which lack the property of meet-continuity, then the category **CONT** of continuous domain and Scott continuous maps is a proper full subcategory of the category **QCONT** of quasicontinuous domains. The problem what the relation is between these two categories, therefore, is interesting and is worthy of being investigated. We give an example to show that **CONT** is not a reflective full subcategory of **QCONT**. When we use the category (**CSLAT**) of continuous semilattices and Scott continuous maps preserving non-empty finite infima instead of **CONT**, the functor of the Smyth powerdomain is proved to be left adjoint to the inclusion functor from **CSL** to **QCONT**. It is well-known that the Scott topology of a continuous domain D has a basis consisting of Scott-open filters. This implies to that the lattice $\Gamma(D)$ of Scott closed sets has enough prime elements, i.e., is a spatial frame. However, for a general dcpo, the Scott topology is usually short of this point. The problem of under what suitable conditions the Scott topology of a dcpo D has a basis of open filters or the topology generated by the Scott-open filters (denoted by $\bar{\sigma}(D)$) is T_0 is unknown. A simple result shows that a dcpo such that its Scott topology has a basis of open filters or it is a semilattice with a T_0 $\bar{\sigma}(D)$-topology must be meet-continuous. Moreover, under appropriate conditions, the category of dcpos with T_0 $\bar{\sigma}(D)$-topology and with Scott continuous maps as morphsims forms a cartesian closed category.

Recall some basic notations for domain, topology and category theories (for details see [1, 3, 8]). A partially ordered set P is said to be directed-complete and is called a dcpo if every directed set D of P has a supremum, denoted $\bigvee D$. A pointed dcpo is one with a bottom element (written $\perp$). For every set $A \subseteq P$, denoted by $\downarrow A = \{b \in P : \exists a \in A, b \leq a\}$ and $\uparrow A = \{b \in P : \exists a \in A, a \leq b\}$. Let P be a dcpo. For $x, y \in P$, we say that x is way below y, written $x \ll y$, if $y \leq \vee D$ implies that there is some $d \in D$ with $x \leq d$ for every directed set $D \subseteq P$. If $x \ll x$ then x is called compact in P. Let $\twoheaddownarrow x = \{y \in P : y \ll x\}$, $k(x) = \{a \in P : a \ll a \leq x\}$ and $\twoheaduparrow x = \{y \in P : x \ll y\}$. A dcpo P is called a continuous (resp., algebraic) dcpo if $\twoheaddownarrow x$ (resp., $k(x)$) is directed and $x = \bigvee \twoheaddownarrow x$ (resp., $x = \bigvee k(x)$) for every $x \in P$. In this case, P is also called a continuous (resp., an algebraic) domain.

Definition 1.1. Let P be a dcpo. A subset U is Scott-open if $U = \uparrow U$ and $\bigvee D \in U$ implies $D \cap U \neq \emptyset$ for every directed set D. The Scott-open sets form a topology called the Scott topology, denoted $\sigma(P)$. Let $w(P) = \{P \backslash \uparrow F : F \subseteq_{fin} P\}$. The topology generated by taken $\sigma(P) \vee w(P)$ as the base is called the Lawson topology, denoted by $\lambda(P)$.

Let P, Q be dcpo's and $f : P \longrightarrow Q$ a function. f is called Scott continuous if it is monotone and preserves sups of all directed sets of P.

Definition 1.2. [6, 7] A dcpo P is called a quasicontinuous domain if for any $x \in P$, $U \in \sigma(P)$, $x \in U$ implies that there exists a finite subset $F \subseteq P$ such that

$$x \in (\uparrow F)^\circ \subseteq \uparrow F \subseteq U,$$

where $(\uparrow F)^\circ$ means the interior of $\uparrow F$ via the Scott topology.

Remark 1.1. (1.2) is equivalent to the original definition of Gierz, Lawson and Stralka in [4].

Lemma 1.1. *[4] Every quasicontinuous domain is locally compact sober with the Scott topology and Hausdorff with the Lawson topology.*

Next we give the definition of a reflective subcategory.

Definition 1.3. [8] Let $\mathcal{C}$ be a category. A subcategory $\mathcal{A}$ of $\mathcal{C}$ is called reflective to $\mathcal{C}$ if the inclusion functor $E : \mathcal{A} \longrightarrow \mathcal{C}$ has a left adjunction, i.e., for each object $C \in \mathcal{C}$, there exist a object $A \in \mathcal{A}$ and a morphism $e : C \longrightarrow A$ in $\mathcal{C}$ such that for any object $B \in \mathcal{A}$ and a morphism $f : C \longrightarrow B$ in $\mathcal{C}$ there is a unique morphism $\bar{f} : A \longrightarrow B$ in $\mathcal{A}$ with $f = \bar{f} \circ e$. e is called a universal map.

In this paper, the category of quasicontinuous (resp., continuous) domains with Scott continuous maps is denoted by **QCONT** (resp., **CONT**).

2. Basic properties of meet-continuous dcpos

Definition 2.1. Let L be a poset. L is call a directed complete semilattice (dcs-lattice, in short) if L is a dcpo and for all $a, b \in L$, $a \wedge b$ exists. L is called a meet-continuous semilattice if L is a dcs-lattice and for $x \in L$ and a directed set $D \subseteq L$,

$$\text{(MCL)} \quad x \wedge \bigvee D = \bigvee_{d \in D} x \wedge d.$$

Definition 2.2. Let P be a dcpo. P is called meet-continuous (mc-dcpo, in short) if for $x \in P$ and a directed set $D \subseteq P$, the following condition holds:

$$\text{(MCD)} \quad x \leq \bigvee D \Rightarrow x \in \mathrm{cl}_\sigma(\downarrow x \cap \downarrow D).$$

Remark 2.1. The above definition was first introduced by Hui Kou [13] in 1998.

Proposition 2.1. *For a dcpo P and a set $A \subseteq P$, $\mathrm{cl}_\sigma A = \downarrow x$ implies $\bigvee A = x$. Hence, if P is meet-continuous and $D \subseteq P$ is directed with $x \leq \bigvee D$, then $x = \bigvee(\downarrow x \cap \downarrow D)$.*

Proof. Given $y \in P$ with $x \in L \backslash \downarrow y$, we have $A \cap L \backslash \downarrow y \neq \emptyset$ since $\mathrm{cl}_\sigma A = \downarrow x$. This means y is not an upper bound of A whenever $x \not\leq y$. Hence, $\bigvee A = x$. □

Proposition 2.2. *(1) Every continuous dcpo is meet-continuous.*
(2) For a dcs-lattice L, (MCL) is equal to (MCD).

Lemma 2.3. *Let P be a mc-dcpo. Then the following hold:*

1 $\uparrow(U \cap \downarrow x) \in \sigma(P)$ for a Scott-open $U \in \sigma(P)$ and $x \in U$.

2 $\uparrow U \in \sigma(P)$ for any Lawson open $U \in \lambda(P)$.

3 If $X = \uparrow X$ for $X \subseteq P$, then $\mathrm{int}_\sigma X = \mathrm{int}_\lambda X$.

4 If $X = \downarrow X$ for $X \subseteq P$, then $\mathrm{cl}_\sigma X = \mathrm{cl}_\lambda X$.

Proof. 1. Suppose $\bigvee D \in \uparrow(U \cap \downarrow x)$, where $U \in \sigma(P)$, $x \in U$ and $D \subseteq P$ is directed. Then there exists $y \in U \cap \downarrow x$ such that $y \leq \bigvee D$. Since P is a mc-dcpo, we have $y \in \mathrm{cl}_\sigma(\downarrow y \cap \downarrow D)$, i.e. $U \cap \downarrow y \cap \downarrow D \neq \emptyset$. Note that since $y \in U \cap \downarrow x$, it follows $U \cap \downarrow x \cap \downarrow D \neq \emptyset$. Hence, $D \cap \uparrow(U \cap \downarrow x) \neq \emptyset$.

2. Suppose $\bigvee D \in \uparrow U$, where $U \in \lambda(P)$ and $D \subseteq P$ is directed. Then we can find $x \in U$ with $x \leq \bigvee D$. From the definition of the Lawson topology, there is a Scott-open $V \in \sigma(P)$ and a finite set $F \subseteq P$ such that $x \in V \backslash \uparrow F \subseteq U$. Note that since P is a mc-dcpo, it follows that $x \in \mathrm{cl}_\sigma(\downarrow x \cap \downarrow D)$, i.e., $V \cap \downarrow x \cap \downarrow D \neq \emptyset$. Since $x \notin \uparrow F$, we have $(V \backslash \uparrow F) \cap \downarrow x \cap \downarrow D = V \cap \downarrow x \cap \downarrow D \neq \emptyset$. Hence, $U \cap \downarrow x \cap \downarrow D \neq \emptyset$, i.e, $\uparrow U \cap D \neq \emptyset$.

3. Suppose $X = \uparrow X$ for $X \subseteq P$. Then $\mathrm{int}_\sigma X \subseteq \mathrm{int}_\lambda X$ since $\sigma(P) \subseteq \lambda(P)$. From (2), $\mathrm{int}_\lambda X \subseteq \uparrow\mathrm{int}_\lambda X = \mathrm{int}_\sigma(\uparrow\mathrm{int}_\lambda X) \subseteq \mathrm{int}_\sigma \uparrow X = \mathrm{int}_\sigma X$. Hence, $\mathrm{int}_\sigma X = \mathrm{int}_\lambda X$.

4. It can be show similarly as (3). □

Theorem 2.4. *Let P be a dcpo. Then the following conditions are equivalent:*

1 P is a mc-dcpo;

2 $\sigma(P)^{\mathrm{op}}$ is a complete Heyting algebra.

Proof. (1)⇒(2) Let $\Gamma(P) = \{A \subseteq P : P \backslash A \in \sigma(P)\}$ be the Scott closed set lattice (ordered by set-theoretic inclusion), then $\sigma(P)^{\mathrm{op}} \cong \Gamma(P)$. Let $\{A_i : i \in I\} \subseteq \Gamma(P)$ be a family of Scott closed sets such that $K \subseteq \bigvee_{i \in I} A_i$. Since $\bigvee_{i \in I} A_i = \mathrm{cl}_\sigma \bigcup_{i \in I} A_i$, it follows $K \subseteq \mathrm{cl}_\sigma \bigcup_{i \in I} A_i$. Assume that there exists

$$x \in K,\ x \notin \bigvee_{i \in I}(K \cap A_i).$$

Note that since $\bigvee_{i \in I}(K \cap A_i) = \mathrm{cl}_\sigma \bigcup_{i \in I} K \cap A_i$, we can find a Scott-open neighborhood U of x such that

$$U \cap \bigcup_{i \in I}(K \cap A_i) = U \cap K \cap \bigcup_{i \in I} A_i = \emptyset.$$

From condition (1) and Lemma 2.5 (1), $\uparrow(U \cap \downarrow x) \in \sigma(P)$. Hence we have

$$(\bigcup_{i\in I} A_i) \cap \uparrow(U \cap \downarrow x) \neq \emptyset$$

because $x \in \uparrow(U \cap \downarrow x)$ and $x \in K \subseteq \mathrm{cl}_\sigma \bigcup_{i\in I} A_i$. It follows that there exists $y \in P$ and $i_0 \in I$ such that $y \in A_{i_0} \cap \uparrow(U \cap \downarrow x)$. This means that there exists $a \in U$ such that $a \in \downarrow x \cap \downarrow y$. Note that since both K and A_{i_0} are Scott closed sets, then it follows $a \in U \cap K \cap A_{i_0} \subseteq U \cap K \cap \bigcup_{i\in I} A_i$. This is a contradiction. Hence $K = \bigvee_{i\in I}(K \cap A_i)$. From proposition 2.2, $\Gamma(P)$ is a meet-continuous lattice, hence is a complete Heyting algebra.

(2)⇒(1) Suppose $x \leq \bigvee D$ for $x \in P$ and a directed set $D \subseteq P$. Then $\{\downarrow d : d \in D\}$ is a directed set of the Scott closed set lattice $\Gamma(P)$. Since each Scott closed set is closed under the supremum of its directed sets, it follows $\bigvee D \in \mathrm{cl}_\sigma \downarrow D$, i.e., $x \in \mathrm{cl}_\sigma \downarrow D$. Since $\bigvee_{d\in D} \downarrow d = \mathrm{cl}_\sigma \downarrow D$, we have $\downarrow x \subseteq \bigvee_{d\in D} \downarrow d$. It follows from condition (2),

$$\downarrow x = \bigvee_{d\in D} (\downarrow x \cap \downarrow d) = \mathrm{cl}_\sigma \bigcup_{d\in D} (\downarrow x \cap \downarrow d) = \mathrm{cl}_\sigma(\downarrow x \cap \downarrow D).$$

This means $x \in \mathrm{cl}_\sigma(\downarrow x \cap \downarrow D)$, i.e., P is a mc-dcpo. □

Note that $\sigma(P)$ is a hypercontinuous lattice and $\Gamma(P)$, the lattice of Scott closed sets, is a generalized continuous lattice when P is quasicontinuous [4]. So if P is also meet-continuous, then from Theorem 2.6, $\Gamma(P)$ is continuous. Therefore, $\sigma(P)$ is a completely distributive lattice. Note that since (P, σ) is isomorphic to the spectrum of $\sigma(P)$, then we have

Theorem 2.5. *A quasicontinuous dcpo P is continuous if and only if it is a mc-dcpo.*

3. Dcpos with the Hausdorff Lawson topology

Generally, the function space of two continuous dcpos may not be continuous. But we can show that it may admit a Hausdorff Lawson topology.

Proposition 3.1. *Let P, E be continuous pointed dcpos. Then the Lawson topology on $[P \longrightarrow E]$ is Hausdorff.*

Proof. Suppose $f \not\leq g$ for $f, g \in [P \longrightarrow E]$. Then there exists $a \in P$ such that $f(a) \not\leq g(a)$. Note that since P, E are continuous, then there are $b, \bar{b} \in P$ and $c \in E$ such that

$$b \ll \bar{b} \ll a,\ c \ll f(b),\ c \not\leq g(a).$$

For $b \ll \bar{b}$ and $c \ll f(b)$, we can find two countable infinite sets $\{b_i\}_{i\in N} \subseteq P$ and $\{c_i\}_{i\in N}$ such that

$$b \ll b_1 \ll \cdots \ll b_n \ll b_{n+1} \ll \cdots \ll \bar{b} \ll a,$$

$$c \ll \cdots \ll c_{n+1} \ll c_n \ll \cdots \ll c_1 \ll f(b).$$

For each $i \in N$, define a map $f_i : P \longrightarrow E$ as follows:

$$f_i(x) = \begin{cases} c_i, & b_i \ll x, \\ \bot, & \text{otherwise.} \end{cases}$$

It is easily see that f_n is Scott continuous and $f_n \leq f$ for all $n \in N$. We claim $f_{n+1} \ll f_n$ for each $n \in N$. Suppose that $D \subseteq [P \longrightarrow E]$ is directed with $f_n \leq \bigvee D$. Since $b_n \ll b_{n+1}$, we have $c_n = f_n(b_{n+1}) \leq (\bigvee D)(b_{n+1})$. Note that since $c_{n+1} \ll c_n$, there is a function $d \in D$ such that $c_{n+1} \leq d(b_{n+1})$. It follows $f_{n+1} \leq d$ from the definition of f_{n+1}. Set

$$\mathcal{U} = \bigcup\{\uparrow f_n : n \in N\},$$

then $\mathcal{U}$ is Scott-open in $[P \longrightarrow E]$ with $f \in \mathcal{U}$. Next we define a function $\bar{f} : P \longrightarrow E$ the taking value c when $\bar{b} \ll x$, $\bot$, otherwise. Since $b_n \ll \bar{b}$ and $c \ll c_n$, it follows $\bar{f} \leq f_n$ for all $n \in N$. Obviously $\bar{f}$ is also Scott continuous with $\bar{f} \not\leq g$. Hence, the set $\mathcal{V} = [P \to E]\backslash\uparrow\bar{f}$ is a Lawson open neighborhood of g and $U \cap V = \emptyset$. This implies that $[P \longrightarrow E]$ is Hausdorff in its Lawson topology.

□

The following result was proved by A. Jung (see [10, 11]).

Theorem 3.2. *If $[P \longrightarrow E]$ is continuous for continuous pointed dcpos P, E, then either P is Lawson compact or E is an L-domain.*

We give an example to show that a dcpo with Hausdorff Lawson topology need not be quasicontinuous even if it is meet-continuous, which is clearly different from the case of complete lattices.

Example 3.3. *Let $P_N = N \cup \{a, b, \bot\}$, where N is the natural number set, ordered by:*

- *$m \sqsubseteq n$ if and only if $n \leq m$ for $m, n \in N$;*
- *$a, b \sqsubseteq n$ for all $n \in N$ and $\bot \sqsubseteq x$ for all $x \in P$;*
- *$a \not\sqsubseteq b$ and $b \not\sqsubseteq a$.*

Then $(P_N, \sqsubseteq)$ is an algebraic pointed dcpo. Hence from Proposition 3.1, $[P_N \longrightarrow P_N]$ has Hausdorff Lawson topology. We claim that $[P_N \longrightarrow$

$P_N]$ is not quasicontinuous. Assume that $[P_N \longrightarrow P_N]$ is quasicontinuous. Note that since P is also a dcs-lattice, then $[P_N \longrightarrow P_N]$ is a dcs-lattice. It follows that $[P_N \longrightarrow P_N]$ is meet-continuous. Hence from Theorem 2.7, $[P_N \longrightarrow P_N]$ is continuous. This is a contradiction from Theorem 3.2, because P_N is neither Lawson compact nor an L-domain (see [1]).

4. Adjunctions between quasicontinuous domains and continuous domains

Since there exist quasicontinuous dcpos which lack the property of meet-continuity, the category **CONT** of continuous domain and Scott continuous maps is a proper full subcategory of the category **QCONT** of quasicontinuous domains. The problem of the relation between these two categories, therefore, is interesting and is worthy of being investigated.

At first, let us consider an example.

Let $N^* = N \cup \{\infty\}$, where N is the set of natural numbers. Define

$$P = (N^* \times \{\infty\}) \cup (\{\infty\} \times N^*),$$

ordered as follows: $\forall m, n \in N$,

- $(m, \infty) \sqsubseteq (n, \infty)$ if and only if $m \leq n$;
- $(\infty, m) \sqsubseteq (\infty, n)$ if and only if $m \leq n$;
- $(m, \infty) \not\sqsubseteq (\infty, n)$, $(\infty, n) \not\sqsubseteq (m, \infty)$;
- $(m, \infty), (\infty, n) \sqsubseteq (\infty, \infty)$.

One can easily see that $(P, \sqsubseteq)$ is not a continuous dcpo, because it is not meet-continous. Suppose that $U \subseteq P$ is a nonempty Scott-open set. Since $(\infty, \infty) \in U$, there exist $m, n \in N$ such that $U = \uparrow\{(m, \infty), (\infty, n)\}$. Hence, every nonempty Scott-open set of P is also a finitary set. It follows from Definition 1.2 that $(P, \sqsubseteq)$ is a quasicontinuous domain.

Theorem 4.1. **CONT** *is not a reflective full subcategory of* **QCONT**.

Proof. Assume that **CONT** is a reflective full subcategory of **QCONT**. Then there is a continuous dcpo E and a Scott continuous map $e : P \longrightarrow E$ such that e is a universal map in **QCONT**, where $P = N^* \times \{\infty\} \cup \{\infty\} \times N^*$ is the dcpo constructed as above. Let

$$\bar{N} = (N^* \times N^*) \backslash \{(n, n) : n \in N\}$$

ordered as follows: $\forall (m_1, n_1), (m_2, n_2) \in \bar{N}$,

$$(m_1, n_1) \leq (m_2, n_2) \text{ iff } m_1 \leq m_2 \text{ and } n_1 \leq n_2,$$

where $m \leq \infty$ for all $m \in N^*$. Then $(\bar{N}, \leq)$ is a algebraic dcpo with $K(\bar{N}) = \{(m,n) : m, n \in N \ \& \ m \neq n\}$ and $P \subseteq \bar{N}$. Define a map $f : P \longrightarrow \bar{N}$ as follows: $\forall n \in N^*$,

$$f(n, \infty) = (n, \infty), \ f(\infty, n) = (\infty, n).$$

One can see that f is Scott continuous. Hence, from the universality of e there exists a unique Scott continuous map $\bar{f} : E \longrightarrow \bar{N}$ such that $f = \bar{f} \circ e$. Now define a map $g : \bar{N} \longrightarrow \bar{N}$ as follows: $\forall (m,n) \in \bar{N}$,

$$g(m,n) = \begin{cases} (m, n-1), & \text{if } m \neq \infty \neq n \text{ and } m+1 < n, \\ (m-1, n), & \text{if } m \neq \infty \neq n \text{ and } n+1 < m, \\ (m,n), & \text{otherwise.} \end{cases}$$

Then g is well defined and $g \circ \bar{f} \circ e(m,n) = g \circ f(m,n) = g(m,n) = (m,n) = f(m,n)$ for each $(m,n) \in P$, i.e., $f = g \circ \bar{f} \circ e$.

(1) g is a Scott continuous function.

For all $(m,n) \in \bar{N}$, $g(m,n) \leq (m,n)$ from the definition of g. Suppose $(m,n), (r,s) \in \bar{N}$ with $(m,n) \leq (r,s)$. Let

$$\begin{aligned} N_1 &= \{(m,n) : m, n \in N \text{ and } |m-n| = 1\}, \\ N_2 &= \{(m,n) : m, n \in N \text{and } m+1 < n\}, \\ N_3 &= \{(m,n) : m, n \in N \text{and } n+1 < m\}. \end{aligned}$$

Then $\bar{N} = P \cup N_1 \cup N_2 \cup N_3$. If $(r,s) \in P \cup N_1$, then $g(m,n) \leq (m,n) \leq (r,s) = g(r,s)$. If $(r,s) \in N_2$, then $(m,n) \notin P$. When $(m,n) \in N_1$, we have $m = n+1$ or $n = m+1$. It follows $m = n+1 \leq r < s-1$ or $n = m+1 \leq r+1 \leq s-1$, i.e., $g(m,n) = (m,n) \leq (r, s-1) = g(r,s)$. One can similarly prove $g(m,n) \leq g(r,s)$ for the other cases. Hence, g preserves the order. Suppose that $D \subseteq \bar{N}$ is directed set. If $\bigvee D = (\infty, \infty)$, then $g(\bigvee D) = (\infty, \infty)$ and for any $m, n \in N$ there exists $(r,s) \in D$ such that $(m,n) \leq g(r,s)$; thus $g(\bigvee D) = \bigvee g(D)$. If $\bigvee D = (m, \infty)$ for $m \in N$, then $\bar{D} = \{(r,s) \in D : r = m\}$ is directed and $\bigvee \bar{D} = \bigvee D$. For each $n \in N$, we can also find $(m,s) \in \bar{D}$ with $n \leq s$, i.e., $g(\bigvee D) = \bigvee g(D)$. If $\bigvee D \in N_1 \cup N_2 \cup N_3$, then D is finite. It follows obviously that $g(\bigvee D) = \bigvee g(D)$. Hence, g is a Scott continuous function.

(2) $g \circ \bar{f} \neq f$.

Let $m \in N$, then $e(m, \infty) \in E$. Since E is a continuous dcpo, $e(m, \infty) = \bigvee \{k \in E : k \ll e(m, \infty)\}$. Since $\bar{f}$ is Scott continuous and $f = \bar{f} \circ e$, we have

$$\begin{aligned} (m,\infty) &= f(m,\infty) \text{ (by the above definition of } f) \\ &= \bar{f} \circ e(m,\infty) \\ &= \bar{f}(\bigvee\{k \in E : k \ll e(m,\infty)\}) \\ &= \bigvee\{\bar{f}(k) : k \ll e(m,\infty)\}. \quad (*) \end{aligned}$$

For each $k \ll e(m,\infty)$, let $\bar{f}(k) = (a_k, b_k)$, then $a_k \leq m$. Notice that since $e(m,\infty) \leq e(\infty,\infty) = e(\bigvee\{(\infty,n) : n \in N\}) = \bigvee\{e(\infty,n) : n \in N\}$, then for each $k \ll e(m,\infty)$ there exists $n \in N$ such that $k \leq e(\infty,n)$; thus

$$\bar{f}(k) = (a_k, b_k) \leq \bar{f} \circ e(\infty,n) = f(\infty,n) = (\infty,n).$$

This shows that for all $k \ll e(m,\infty)$,

$$(a_k, b_k) \in N \times N, \ a_k \leq m.$$

Hence, from the equation $(*)$, there exists $k_0 \ll e(m,\infty)$ such that $a_{k_0} \leq m < m+1 < b_{k_0} < \infty$. It follows from the definition of g that

$$g \circ \bar{f}(k_0) = g(a_{k_0}, b_{k_0}) = (a_{k_0}, b_{k_0} - 1) \neq (a_{k_0}, b_{k_0}) = \bar{f}(k_0).$$

This is a contradiction to the uniqueness of $\bar{f}$. Hence, P has no universal object in **CONT**, i.e., **CONT** is not a reflective full subcategory of **QCONT**. □

From Theorem 4.1, **CONT** is not a reflective subcategory of **QCONT**. So we need to investigate more special categories than **CONT**.

Definition 4.1. Let $(X, \Omega(X))$ be a topological space. For a set $A \subseteq X$, let

$$Sat(A) = \bigcap\{U : A \subseteq U \in \Omega(X)\}.$$

A is called a saturated set if $A = Sat(A)$. It is easy to show that for any $U \in \Omega(X)$, $Sat(A) \subseteq U$ if and only if $A \subseteq U$, $Sat(A)$ is compact if and only if A is compact. In particular, if X is a dcpo with the Scott topology $\sigma(X)$ then every Scott compact saturated set is exactly a Scott compact upper set. Let $S(X)$ be the family of all nonempty non-empty saturated compact set of X, ordered by: $A \sqsubseteq B$ if and only if $B \subseteq A$.

The following is the well-known Keimel-Paseka Lemma.

Lemma 4.2. *[12] For a sober space X, $(S(X), \sqsubseteq)$ is a dcpo and for each filtered family $\mathcal{F}$ of $S(X)$, $\bigcap \mathcal{F} \subseteq U$ implies that there exists some $K \in \mathcal{F}$ such that $K \subseteq U$ for $U \in \Omega(X)$.*

In this case, the structure $(S(X), \sqsubseteq)$ is called the Smyth powerdomain of X (see [6, 7, 14, 23]).

Note that each quasicontinuous domain is sober locally compact via the Scott topology.

Lemma 4.3. *[6] For a quasicontinuous domain, $S(P)$ is a continuous semilattice such that:*

- *For $C_1, C_2 \in S(P)$, $C_1 \ll C_2$ if and only if there exists a finite set $F \subseteq P$ such that $C_2 \subseteq (\uparrow F)^\circ \subseteq C_1$,*
- *For each $C \in S(P)$, $C = \bigcap\{\uparrow F : F \subseteq_{fin} P \;\&\; C \subseteq (\uparrow F)^\circ\}$, where the right family is filtered and $(\uparrow F)^\circ$ is the Scott interior,*
- *For $C_1, C_2 \in S(P)$, $C_1 \wedge C_2 = C_1 \cup C_2$.*

Denote by **CSL** the category of continuous semilattices with Scott continuous functions which preserve nonempty finite infima as morphisms.

Theorem 4.4. *The category of continuous semilattices (**CSL**) is a reflective subcategory of the category of quasicontinuous domains(**QCONT**), i.e., S is the left adjunction of E, where E is the inclusion functor from **CSL** to **QCONT**.*

Proof. Its proof can be found in [15]. □

Definition 4.2. A quasicontinuous domain P is called coherent provided P is Scott quasicompact and

$$\forall C_1, C_2 \in S(P),\ C_1 \cap C_2 \neq \emptyset \Rightarrow C_1 \cap C_2 \in S(P).$$

Remark 4.1. If P is viewed as a topological space via its Scott topology, P is a coherent topological space (see [1]). A number of important domains, such as FS-domains, bifinite domains, Scott domains (see [1]), etc., are coherent.

Proposition 4.5. *Let P be a quasicontinuous domain with a least element $\perp$. Then the following conditions are equivalent:*

1 The Lawson topology $\lambda(P)$ on P is compact;

2 For $x, y \in P$, $\uparrow x \cap \uparrow y$ is Lawson compact;

3 P is coherent;

4 $S(P)$ is a bounded complete domain.

Proof. $1 \Rightarrow 2$: Obviously.

$2 \Rightarrow 3$: Since P has a bottom $\perp$, P is Scott compact. Suppose $C_1 \cap C_2 \neq \emptyset$ for $C_1, C_2 \in S(P)$. Then from Lemma 4.4,

$$C_1 \cap C_2 = \bigcap\{\uparrow F_1 \cap \uparrow F_2 : F_1, F_2 \subseteq_{fin} P \,\&\, C_1 \subseteq (\uparrow F_1)^\circ,\ C_2 \subseteq (\uparrow F_2)^\circ\} \in S(P).$$

$3 \Rightarrow 4$: This follows directly from Definition 4.6.
$4 \Rightarrow 1$: We use the Alexander Lemma: a space is compact if every open cover consisting of subbasic open sets contains a finite subcover. From condition 4, one can easily see that for all $C_1, C_2 \in S(P)$, $C_1 \cap C_2 \neq \emptyset$ implies $C_1 \vee_S C_2 = C_1 \cap C_2$. Let

$$\mathcal{U} = \{U_j \in \sigma(P) : j \in J\} \bigcup \{P \backslash \uparrow x_k : k \in K\}$$

be an open cover of P. Since $\bigcup\{P\backslash \uparrow x_k : k \in K\} = P \backslash \bigcap_{k\in K} \uparrow x_k$, we have

$$\bigcap\{\uparrow x_k : k \in K\} \subseteq \bigcup\{U_j : j \in J\}.$$

If $\bigcap\{\uparrow x_k : k \in K\} = \emptyset$, then $P = \bigcup\{P\backslash\uparrow x_k : k \in K\}$. We claim that there exists a nonempty finite set $F \subseteq K$ such that $\bigcap\{\uparrow x_k : k \in F\} = \emptyset$. Assume that for any nonempty finite set $F \subseteq K$, $\bigcap\{\uparrow x_k : k \in F\} \neq \emptyset$. Then $\mathcal{F} = \{\bigcap_{k\in F} \uparrow x_k : \emptyset \neq F \subseteq_{fin} K\}$ is a filtered family of nonempty Scott compact sets. Hence $\bigcap\{\uparrow x_k : k \in K\} = \bigcap \mathcal{F} \neq \emptyset$, a contradiction. Let F_0 be a nonempty subset of K with $\bigcap_{k\in F_0} \uparrow x_k = \emptyset$. Then $P = \bigcup\{P\backslash\uparrow x_k : k \in F_0\}$. If $\bigcap\{\uparrow x_k : k \in K\} \neq \emptyset$, then $\mathcal{F} = \{\bigcap_{k\in F} \uparrow x_k : \emptyset \neq F \subseteq_{fin} K\}$ is a filtered family of nonempty Scott compact upper sets. Note that since $\bigcap \mathcal{F} =\subseteq \bigcup\{U_j : j \in J\}$, then from lemma 3.1 there exists a nonempty finite set $F \subseteq K$ such that $\bigcap_{k\in F} \uparrow x_k \subseteq \bigcup\{U_j : j \in J\}$. Since $\bigcap\{\uparrow x_k : k \in F\}$ is Scott compact, there is a finite set $F' \subseteq J$ such that $\{\uparrow x_k : k \in F\} \subseteq \bigcup\{U_s : s \in F'\}$. Let $\mathcal{U}_\prime = \{U_s : s \in F'\} \cup \{P\backslash\uparrow x_k : k \in F\}$, then $\mathcal{U}_\prime$ is a finite subcover of $\mathcal{U}$. □

Denote by **QCCOH**$_\perp$ the category of coherent pointed quasicontinuous domains with Scott continuous morphisms, and **CBD** the category of bounded complete continuous domains with Scott continuous morphisms which preserve nonempty finite meets. Then from Theorem 4.5 and Proposition 4.7 we have the following result.

Theorem 4.6. **CBD** *is a reflective subcategory of* **QCCOH**$_\perp$.

5. Scott-open filters

It is well-known that the Scott topology of a continuous domain D has a basis consisting of Scott-open filters. This implies that the lattice $\sigma(D)$ of Scott-open sets has enough coprime elements. However, for a general dcpo, the Scott topology is usually short of this point. The problem that under what suitable conditions the Scott topology of a dcpo has a basis of open filters or the topology generated by the Scott-open filters is T_0 is unknown (see [17, Problem 527]). In the following, we will investigate this question.

Definition 5.1. Let P be a dcpo. A set $F \subseteq P$ is filtered if $F \neq \emptyset$ and for all $x, y \in F$, there is $z \in F$ such that $z \leq x, y$. In this case, if $F = \uparrow F$, then F is called a filter. We say that a poset with a topology has small open filtered sets if and only if each point has a neighborhood basis of open filtered sets.

Proposition 5.1. *For a dcpo P, the following are equivalent:*

1 The Scott topology $\sigma(P)$ has a basis of open filters;

2 P is meet-continuous and the Lawson topology $\lambda(P)$ has a small open filtered sets.

Proof. (1)$\Rightarrow$(2) Suppose $x \leq \bigvee D$ for $x \in P$ and for a directed set $D \subseteq P$. Then for any Scott-open filter V, $x \in V$ means $D \cap V \neq \emptyset$. Pick $a \in D \cap V$. Note that since V is a filter, then there exists $b \in V$ such that $b \leq a$ and $b \leq x$, i.e., $b \in \downarrow x \cap \downarrow D \cap V$. Hence from condition (1), $x \in \mathrm{cl}_\sigma(\downarrow x \cap \downarrow D)$, i.e, P is meet-continuous. Suppose that $W \subseteq P$ is a Lawson open neighborhood of x. Then there are a Scott-open filter U and a finite set F such that $x \in U \backslash \uparrow F$. Obviously, $U \backslash \uparrow F$ is a Lawson open filtered set.

(2)$\Rightarrow$(1) Suppose $x \in U$ for $x \in P$ and $U \in \sigma(P)$. Since $\sigma(P) \subseteq \lambda(P)$, there exists a Lawson open filtered set V such that $x \in V \subseteq U$. Note that since P is meet-continuous, then from Lemma 2.5, $\uparrow V$ is Scott-open. Therefor, the Scott topology $\sigma(P)$ has a basis of open filters. □

Let P be a dcpo. Set

$$SF(P) = \{V \in \sigma(P) : V \text{ is a filter}\}.$$

In order to ensure $SF(P) \neq \emptyset$, we assume that P is a $(\wedge)$ semilattice (i.e., dcs-lattice) from now on. In this case, P itself is a Scott-open filter and $SF(P)$ forms a topological basis. Denote $\bar{\sigma}(P)$ the topology generated by $SF(P)$, and $\bar{\lambda}(P) = \bar{\sigma}(P) \vee w(P)$, where $w(P)$ means the lower topology. Obviously we have $\bar{\sigma}(P) \subseteq \sigma(P)$ and $\bar{\sigma}(P) \subseteq \bar{\lambda}(P) \subseteq \lambda(P)$.

Proposition 5.2. *For a dcs-lattice P, if the topology $\bar{\sigma}(P)$ is T_0, then P is meet-continuous.*

Proof. Suppose $x \leq \bigvee D$ and $\bigvee_{d \in D} x \wedge d < x$ for $x \in P$ and for a directed set $D \subseteq P$. Since $\bar{\sigma}(P)$ is T_0 and all Scott-open sets are upper sets, there exists an open filter $V \in SF(P)$ such that $x \in V \subseteq P \backslash \bigvee xD$. Note that since $x \leq \bigvee D$, we have $D \cap V \neq \emptyset$. Pick $d \in D \cap V$, then $d \wedge x \in V$ because P is a semilattice and V is a filter. This is a contradiction. $\square$

Theorem 5.3. *Let P be a dcs-lattice. Then $\bar{\sigma}(P)$ is T_0 if and only if $\bar{\lambda}(P)$ is T_1.*

Proof. If $\bar{\sigma}(P)$ is T_0, then for all $x, y \in P$, $\mathrm{cl}_{\bar{\sigma}}\{x\} = \mathrm{cl}_{\bar{\sigma}}\{y\}$ implies $x = y$. We claim that for all $x \in P$, $\{x\} = (\mathrm{cl}_{\bar{\lambda}}\{x\}) \cap \uparrow x$. Assume $\{x\} \neq (\mathrm{cl}_{\bar{\lambda}}\{x\}) \cap \uparrow x$ for some $x \in P$. Then there is $y \in (\mathrm{cl}_{\bar{\lambda}}\{x\}) \cap \uparrow x$ such that $x < y$. Note that since $\bar{\sigma}(P) \subseteq \bar{\lambda}(P)$ and every open set in $\bar{\sigma}(P)$ is a upper set, we have $\mathrm{cl}_{\bar{\sigma}}\{x\} \supseteq \mathrm{cl}_{\bar{\lambda}}\{x\} \supseteq \mathrm{cl}_{\bar{\lambda}}\{y\}$ and $\mathrm{cl}_{\bar{\sigma}}\{x\} = \downarrow \mathrm{cl}_{\bar{\sigma}}\{x\}$. It follows $\mathrm{cl}_{\bar{\sigma}}\{x\} \supseteq \mathrm{cl}_{\bar{\sigma}}\{y\} \supseteq \mathrm{cl}_{\bar{\sigma}}\{x\}$, i.e., $\mathrm{cl}_{\bar{\sigma}}\{x\} = \mathrm{cl}_{\bar{\sigma}}\{y\}$. Hence, $x = y$, a contradiction. So $\{x\} = (\mathrm{cl}_{\bar{\lambda}\{x\}}) \cap \uparrow x$, i.e., $\{x\}$ is a $\bar{\lambda}(P)$-closed set.

Conversely, If $\bar{\lambda}(P)$ is T_1, then for all $x \in P$, $\{x\} = \mathrm{cl}_{\bar{\lambda}}\{x\}$. Take $x, y \in P$ with $\mathrm{cl}_{\bar{\sigma}}\{x\} = \mathrm{cl}_{\bar{\sigma}}\{y\}$. Then for any Scott-open filter $V \subseteq P$, $x \in V$ if and only if $y \in V$ if and only if $x \wedge y \in V$. This means $\mathrm{cl}_{\bar{\sigma}}\{x\} = \mathrm{cl}_{\bar{\sigma}}\{x \wedge y\} = \mathrm{cl}_{\bar{\sigma}}\{y\}$. Assume $x \not\leq y$, then $x \wedge y < x$. Since $\bar{\lambda}(P)$ is T_1, there exist a Scott-open filter and a finite set F such that $x \in V \backslash \uparrow F$ and $x \wedge y \notin V \backslash \uparrow F$. Note that since $x \in V$ and $\mathrm{cl}_{\bar{\sigma}}\{x\} = \mathrm{cl}_{\bar{\sigma}}\{x \wedge y\}$, then $x \wedge y \in V$; thus $x \in \uparrow F$. This is impossible for $x \notin \uparrow F$. Hence $x = y$ and the topology $\bar{\sigma}(P)$ is T_0. $\square$

The following is a direct application of the above result.

Theorem 5.4. *A complete lattice L is a continuous lattice if and only if $\bar{\lambda}(L)$ is Hausdorff.*

Next, we consider the Scott-open filters of function spaces.

For dcpos P, E, let $[P \longrightarrow E] = \{f : P \longrightarrow E : f$ is Scott continuous$\}$. Then $[P \longrightarrow E]$ with the pointwise order is a dcpo.

Lemma 5.5. *[3] For dcpos P, E, $[P \longrightarrow E]$ is a meet-continuous dcs-lattice if and only if E is meet-continuous dcs-lattice.*

The following result is crucial for the further discuss.

Lemma 5.6. *Let P be a dcpo and E a meet-continuous dcs-lattice with a top $\top_E$. For a nonempty Scott open set $U \in \sigma[P \longrightarrow E]$ and $a \in P$, define*

$$U(a) = \{f(a) : f \in U\}.$$

Then $U(a)$ is Scott-open in E.

Proof. (1) $U(a) = \uparrow U(a)$.

Take $t \in E$ and $f \in U$ with $f(a) \leq t$. Define a map $g : P \longrightarrow E$ as follows:

$$g(x) = \begin{cases} t, & x \leq a, \\ \top_E, & x \not\leq a. \end{cases}$$

Then g is well defined and monotone. Next we show that g is Scott continuous. Let $D \subseteq E$ be a directed set.

Case 1: If $\bigvee D \leq a$, then $g(\bigvee D) = t = \bigvee_{d \in D} g(d) = \bigvee g(D)$.

Case 2: If $\bigvee D \not\leq a$, then there exists $d \in D$ such that $d \not\leq a$. It follows $g(\bigvee D) = \top_E = g(d) = \bigvee g(D)$.

These two cases show that g is Scott continuous. Clearly, $f \leq g$, hence $t = g(a) \in U(a)$.

(2) $\bigvee D \in U(a)$ implies $D \cap U(a) \neq \emptyset$ for a directed set $D \subseteq E$.

Since $\bigvee D \in U(a)$, there exists $f \in U$ such that $f(a) = \bigvee D$. For each $d \in D$, define a map $g_d : P \longrightarrow E$ as follows: $\forall x \in P$,

$$g_d(x) = \begin{cases} f(s) \wedge d, & x \leq a, \\ \top_E, & x \not\leq a. \end{cases}$$

Then g_d is well defined and monotone. Let $S \subseteq E$ be a directed set.

Case 1: If $\bigvee S \leq a$, then $g_d(\bigvee S) = f(\bigvee S) \wedge d = d \wedge \bigvee f(S) = \bigvee_{s \in S} d \wedge f(s)$ (because of the meet-continuity of E) $= \bigvee_{s \in S} g_d(s) = \bigvee g_d(S)$.

Case 2: If $\bigvee S \not\leq a$, then there is some $s \in S$ such that $s \not\leq a$. Thus, $g_d(\bigvee S) = \top_E = g_d(s) = \bigvee g_d(S)$.

Hence $g_d \in [P \longrightarrow E]$. Set $\mathcal{D} = \{g_d : d \in D\}$, then $\mathcal{D}$ is directed. Pick $x \in P$. If $x \leq a$, then $\bigvee \mathcal{D}(x) = \bigvee_{d \in D} g_d(x) = \bigvee_{d \in D} f(x) \wedge d = f(x) \wedge \bigvee D = f(x) \wedge f(a) = f(x)$. If $s \not\leq a$, then $\bigvee \mathcal{D}(x) = \top_E$. Hence, $f \leq \bigvee \mathcal{D}$. It follows $\mathcal{D} \cap U \neq \emptyset$ since U is Scott-open in $[P \longrightarrow E]$ and $f \in U$. Take $g_{d_0} \in \mathcal{D} \cap U$ with $d_0 \in D$, then $g_{d_0}(a) = f(a) \wedge d_0$. Note that since $U(a) = \uparrow U(a)$ and $f(a) \in U(a)$, it follows that $d_0 \in D \cap U(a)$. The proof is completed.

□

Theorem 5.7. *Let P, E be dcs-lattices with tops $\top_P$ and $\top_E$ respectively. Then $\bar{\sigma}[P \longrightarrow E]$ is T_0 if and only if $\bar{\sigma}(E)$ is T_0.*

Proof. Suppose that $(E, \sigma_F(E))$ is T_0. Then, from Proposition 5.3 and Lemma 5.6, both E and $[P \longrightarrow E]$ are meet-continuous semilattices. If $f \not\leq g$ for $f, g \in [P \longrightarrow E]$, then there exists $x \in P$ such that $f(x) \not\leq g(x)$. Hence, $(f \wedge g)(x) = f(x) \wedge g(x) < f(x)$. Therefore we can find a Scott-open filter $V \subseteq E$ such that

$$f(x) \in V, \ g(x) \wedge f(x) \notin V. \qquad (*)$$

Obviously, $g(x) \notin V$. Set

$$\mathcal{V} = \{h \in [P \longrightarrow E] : \exists y \in V,\ h(x) \geq y\}.$$

Then from $(*)$, $f \in \mathcal{V}$ and $g \notin \mathcal{V}$. Hence, if $\mathcal{V}$ is a Scott-open filter, then $([P \longrightarrow E], \sigma_F[P \longrightarrow E])$ is T_0. Obviously, $\mathcal{V}$ is an upper set, i.e., $\mathcal{V} = \uparrow\mathcal{V}$. Given a directed set $D \subseteq [P \longrightarrow E]$ with $\bigvee D \in \mathcal{V}$, there is some $y \in V$ such that $(\bigvee D)(x) \leq y$. Since $V \in \sigma(E)$ and $(\bigvee D)(x) = \bigvee_{d\in D} d(x)$, there exists $d \in D$ such that $d(x) \in V$. By the definition of $\mathcal{V}$, $d \in \mathcal{V}$, i.e., $\mathcal{V} \in \sigma[P \longrightarrow E]$. Pick $h, k \in \mathcal{V}$. Then there exist $y_1, y_2 \in V$ such that $h(x) \geq y_1$, $k(x) \geq y_2$. Hence, $(h \wedge k)(x) = h(x) \wedge k(x) \geq y_1 \wedge y_2$. Since V is a filter, we have $(h \wedge k)(x) \in V$. By the definition of $\mathcal{V}$, $h \wedge k \in \mathcal{V}$. It follows that $\mathcal{V}$ is a Scott-open filter of $[P \longrightarrow E]$.

Conversely, if $([P \longrightarrow E], \sigma_F[P \longrightarrow E])$ is T_0, then from Lemma 5.6, T is meet-continuous. Take $x, y \in E$ with $x \not\leq y$. Then, $f_x \not\leq f_y$, where $f_x, f_y \in [P \longrightarrow E]$ are constant functions taking values x and y respectively. Hence there exists a Scott-open filter $\mathcal{V} \subseteq [P \longrightarrow E]$ such that

$$f_x \in \mathcal{V},\ f_y \notin \mathcal{V}.$$

Set $V = \mathcal{V}(\top_P)$. Then from Lemma 5.7, V is a Scott-open filter of E and $x \in V$, $y \notin V$. □

Definition 5.2. A dcpo P is called a SFT_0 semilattice if P is a semilattice with a top and its topology $\bar{\sigma}(P)$ is T_0.

Denote by $\mathbf{SFT}_0$ the category of SFT_0 semilattices with Scott continuous maps as morphisms.

Proposition 5.8. $\mathbf{SFT}_0$ *has finite products.*

Proof. It is sufficient to show that the cartesian product $E_1 \times E_2$ of two SFT_0 semilattices also is an SFT_0 semilattice. Obviously, $E_1 \times E_2$ is a dcs-lattice with the largest element $(\top_1, \top_2)$. Let $(x_1, y_1), (x_2, y_2) \in E_1 \times E_2$ with $(x_1, y_1) \not\leq (x_2, y_2)$. We may assume $x_1 \not\leq x_2$. Then there exists a Scott-open filter $V_1 \subseteq E_1$ such that $x_1 \in V_1$, $x_2 \notin V_1$. Set $V = V_1 \times E_2$. One can easily see that V is a Scott-open filter of $E_1 \times E_2$ such that $(x_1, y_1) \in V$, $(x_2, y_2) \notin V$. □

Combining Theorem 5.8 and Proposition 5.10, we have

Theorem 5.9. $\mathbf{SFT}_0$ *is a cartesian closed category.*

Proposition 5.10. *Let P be a dcpo with bottom $\perp$ and E a dcs-lattice with top $\top$. If $[P \longrightarrow E]$ has a Scott-open filter base, then E has a Scott-open filter base.*

Proof. Pick $U \in \sigma(E)$ and $t \in P$ with $t \in U$. Set $[U] = \{f_x : x \in U\}$, where f_x is the function from P into E defined by taking value x for any $y \in P$. Obviously each f_x is a Scott continuous map. We claim that $[U]$ is a Scott open set of $[P \longrightarrow P]$. Given a directed set $D \subseteq [P \longrightarrow P]$ such that $\bigvee D \in \uparrow[U]$, there exists some $x \in U$ such that $f_x \leq \bigvee D$; thus $x \leq \bigvee_{d \in D} d(\bot)$. Since U is a Scott-open neighborhood of x, there exists $d \in D$ such that $d(\bot) \in U$, i.e., $f_{d(\bot)} \in [U]$. Note that since $\bot$ is the least element of P and d is monotone, it follows that $f_{d(\bot)} \leq d$. Hence, $d \in \uparrow[U]$, i.e., $\uparrow[U] \in \sigma[P \longrightarrow E]$. Since the Scott topology of $[P \longrightarrow E]$ has a Scott-open filter base, there exists a Scott-open filter $V \subseteq [P \longrightarrow E]$ such that $f_x \in V \subseteq \uparrow[U]$ for each $x \in U$. Hence,

$$x \in V(\bot) \subseteq (\uparrow[U])(\bot) = \uparrow([U](\bot)) = U,$$

where $V(\bot) = \{f(\bot) : f \in V\}$. Obviously $V(\bot)$ is a filter of E. We also claim that $V(\bot)$ is Scott-open. Since the Scott topology of $[P \longrightarrow E]$ has a Scott-open filter base, $[P \longrightarrow E]$ is a meet-continuous semilattice from Proposition 5.2 and hence E is a meet-continuous semilattice from Lemma 5.6. It follows from Lemma 5.7 that $V(\bot)$ is Scott-open in E. □

6. Concluding remarks

We used the Scott topology to define meet-continuity on general dcpos and indicated the basic relationship among meet-continuity, Hausdorff separation, quasicontinuity, continuity and Scott-open filter bases of dcpos. For a continuous semilattice P, if P is neither an L-domain or Lawson compact, then function space $[P \longrightarrow P]$ exactly provide an example that dcpos with the Hausdorff Lawson topology need not quasicontinuous, which is a clear difference from the case of complete lattice. However, the problem that under which appropriate conditions the Hausdorff separation implies quasicontinuity is open (for the recently results, refer to [18] of Lawson).

Acknowledgements. The authors gratefully acknowledge the referee for his careful reading and offering helpful suggestions and corrections.

References

[1] S. Abramsky and A. Jung, Domain theory, In Handbook of logic in Computer Science Vol. 3, S.Abramsky, D.M.Gabay and E.Maibaum (editors), Spring-Verlag, 1994.

[2] R.M. Amadio and P-L Curien, Domains and Lambda-Calculi, Cambridge University Press, 1998.

[3] G. Gierz et al, A Compendium of Continuous Lattices, Springer-Verlag, 1980.

[4] G. Gierz, J.D. Lawson and A. Stralka, Quasicontinuous posets, Houston Journal of Mathematics, 9(1983) 191–208.

[5] C. Gunter, Semantics of Programming Languages, Structures and Techniques, Foundations of Computing, MIT Press, 1992.

[6] R. Heckmann, Power Domain Constructions, PhD thesis, Universität des Saarlandes, 1990.

[7] R. Heckmann, An upper power domain construction in terms of strongly compact sets, Lecture Notes in Computer Science 598, Springer-Verlag, 1992, 272–293.

[8] H. Herrlich and E. Strecker, Category theory, Heldermann Verlag, Berlin, 1973.

[9] J.R. Isbell, Function spaces and adjoints, Symposia Math. 36 (1975), 317–339.

[10] A. Jung, The classification of continuous domains, In Logic in Computer Science, IEEE Computer Society Press, 1990.

[11] A. Jung, Cartesian Closed Categories of Domains, Volume 66 of CWI Tracts, 1989.

[12] K. Keimel and J. Paseka, A direct proof of the Hofmann-Mislove theorem, Proceedings of the American Mathematical Society, (120)1, 1994, 301–303.

[13] Hui Kou, On some questions in domain theory and locale theory, Doctoral dissertation, Sichuan university, 1998.

[14] Hui Kou, U_k-admitting dcpos need not be sober, In: Domians and Processes, Edited by K. Keimel, G.Q. Zhang, Y.M. Liu and Y.X. Chen, Proceedings of the 1st international Symposiun on Domain Theory, 2001, 41-50.

[15] Hui Kou, Mao-Kang Luo, Adjunctions between the category of quasicontuous domains and its subcategories, Chinese Annals of Mathematics, 23A:5(2002),633-642.

[16] J. Lawson, The versatile continuous order, Lecture Notes in Computer Science 298, Springer-Verlag, 1988, 134–160.

[17] J. Lawson and M. Mislove, Domain theory and topology, In J.Van Mill and G.M.Reed, editors, Open problems in topology, North-Holland, Amsterdam, 1990.

[18] J. Lawson, The upper interval topology, property M, and compactness, Electronic Notes in Theoretical Comp. Sci., 13(1998).

[19] P.T. Lambrinos and B. Papadopoulos, The (strong) Isbell topology and (weakly) continuous Lattices, In: Continuous and Their Applications, R.E.Hoffmann and K.H.Hofmann (editors), Marcel Dekker, 1985, 191–211.

[20] J.H. Liang and K. Keimel, Compact Continuous L-domains, Computers and Mathematics with Applications 38(1999) 81-89.

[21] M. Mislove, Topology, domain theory and theoretical computer science, Topology and its Applications 89(1998) 3-59.

[22] D.S. Scott, Outline of a mathematical theory of computation. In 4th Annual Princeton Conference on Information Sciences and Systems, 1970, 169-176.

[23] M.B. Smyth, Power domains and predicate transformers: A topological view, Lecture Notes in Computer Science 154, Springer-Verlag, 1983, 662–676.

[24] G. Winskel and K. Larsen, Using information systems to solve recursive domain equations effectively, Lecture Notes in Computer Science 173, Spring-Verlag, 1984, 109-130.

Chapter 6

EXTERNAL CHARACTERIZATIONS OF CONTINUOUS SL-DOMAINS

Luoshan Xu*
Department of Mathematics
Yangzhou University
Yangzhou 225002, P.R. China
luoshanxu@yahoo.com

Abstract In this paper a dcpo L is defined to be an sL-domain if all the principal ideals of L are join semilattices. With this concept, it is proved that a dcpo L is a continuous sL-domain if and only if $[X \to L]$ is a continuous dcpo (resp: sL-domain) for all compact and core compact spaces X. This characterization solves Problem 532 posed by J. D. Lawson and M. W. Mislove in the book [J. Van Mill, G. M. Reed (Editors), Open Problems in Topology, Elsevier Sci. Publishers B.V., North-Holland, 1990, pp 351-372]. Via *idl*(L), some other external characterizations of continuous sL-domains are also given.

Keywords: Continuous dcpo; Scott-open set; function space; sL-domain; core compact and compact space

1. Introduction

In [5, Problem 532], Lawson and Mislove posed the following problem:

> Characterize those L for which $[X \to L]$ is a continuous dcpo for all core compact spaces X. A likely candidate is the class of all continuous L-domains L. Does one get the same answer if one restricts to all the compact and core compact spaces?

Liu and Liang in [6, Theorem 2.15] proved that if L is a dcpo , then L is a continuous L-domain if and only if $[X \to L]$ is a continuous dcpo

*Supported by the National Nature Science Foundation of China and by the Foundation from Education Section of Jiangsu Province, China

GQ Zhang et al. (eds.), Domain Theory, Logic and Computation, 137-149.

for all core compact spaces X, solving the first part of the problem. [6, Proposition 2.16] also gave an example of non-L-domain named T_1 (three points, shaped like Λ) such that for all compact and core compact spaces X, $[X \to T_1]$ is a continuous dcpo, solving the second part of the quoted problem negatively.

Now, the question is: for what kind of dcpo L which is more general than L-domains, is $[X \to L]$ continuous for all compact and core compact spaces X? To answer the question, the concept of an sL-domain, a generalization of an L-domain, is introduced. With retraction methods and construction of special topological spaces (resp: special continuous dcpo's), it is proved that a dcpo L is a continuous sL-domain if and only if $[X \to L]$ is a continuous dcpo (resp: sL-domain) for all compact and core compact spaces X. When the poset $idl(L)$ of all ideals of L is concerned, it is also shown that a continuous dcpo L is a continuous sL-domain if and only if $[X \to idl(L)]$ is a continuous dcpo for all compact and core compact spaces X if and only if $idl(L)$ is an algebraic domain. These results can be viewed as not only deeper answers to the second part of the quoted problem, but also an external characterizations of continuous sL-domains.

2. Preliminaries

Let P be a poset, $x, y \in P$, $x \leq y$. The set $\{z \in P : x \leq z \leq y\}$ is called a *closed interval* in P and denoted by $[x, y]$. If $A \subset P$, then the set $\{z \in P : \exists x \in A,\ x \leq z\}$ is called the upper set of A, denoted by $\uparrow A$. When $A = \{a\}$, we write $\uparrow a$ for $\uparrow A$. By a *dcpo*, we mean a poset in which every directed set has the supremum. Say that a dcpo P is bicomplete (see [3, 6]) if P^{op} (P in the reverse order) is also a dcpo. Say that P is an *L-domain* (see [3, 6]) if P is a dcpo and $\forall x \in P$, $\downarrow x$ is a complete lattice.

Let P be a dcpo, $x \in P$. The set $\{a \in P : a \ll x\}$ is denoted by $\twoheaddownarrow x$ and the set $\{b \in P : x \ll b\}$ is denoted by $\twoheaduparrow x$, where $\ll$ is the way-below relation in P. $k \in P$ is said to be *compact*, if $k \ll k$. P is said to be *continuous*, if $\forall x \in P, \twoheaddownarrow x$ is directed and $\vee^{\uparrow}(\twoheaddownarrow x) = x$. P is said to be *algebraic*, if every element is a directed sup of compact elements. If P is continuous, then the *Scott topology* $\sigma(P)$ of P has a base $\{\twoheaduparrow x : x \in P\}$. If P is continuous and $U \neq \emptyset$ is a Scott open set of P, then U in the order of P is also a continuous dcpo. The set of all ideals of P under set inclusion order is denoted by $idl(P)$. Note that $idl(P)$ is an algebraic dcpo if P is a dcpo.

Lemma 2.1. *([3, 6, 11])* Let P be a dcpo. Then P is continuous if and only if $\forall x \in P$, the principal ideal $\downarrow x$ is continuous.

Proposition 2.2. Let P be a dcpo and $P \cup \{\bot\}$ be the dcpo obtained by adding a bottom element $\bot$ to P. Then P is continuous if and only if $P \cup \{\bot\}$ is continuous.

Proof. If P is continuous, then for every element $x \in P \cup \{\bot\}$, $\Downarrow_{(P\cup\{\bot\})} x = \{\bot\} \cup (\Downarrow_P x)$ is directed and has x as its directed join. So, $P \cup \{\bot\}$ is continuous. Conversely, if $P \cup \{\bot\}$ is continuous then P is a Scott-open set of $P \cup \{\bot\}$ and thus is continuous. □

Lemma 2.3. *([3, Theorem 1.37])* Let P be a (continuous) dcpo. If $[P \to P]$ is continuous, then P is bicomplete.

Proposition 2.4. Let L be a continuous dcpo and A a set of compact elements of L. Then $U = \uparrow A$ is a Scott open set of L.

Proof. It is easy to see that $\forall k \in A, \uparrow k = \Uparrow k$ is Scott open in L. So, $U = \bigcup_{k \in A} \Uparrow k = \uparrow A$ is a union of Scott open sets and thus a Scott open set. □

A topological space X is called core compact if the lattice of all its open sets is a continuous lattice. Let X be a topological space and L a dcpo with the Scott topology $\sigma(L)$. Write $[X \to L]$ for the set of all continuous mappings from X to $(L, \sigma(L))$. Then, $[X \to L]$ in the pointwise order is also a dcpo. It is well-known that if P is a continuous dcpo, then its Scott topology $\sigma(P)$ is a completely distributive complete lattice and thus the space $(P, \sigma(P))$ is core compact. This yields

Proposition 2.5. If $P = (\uparrow a) \cup (\uparrow b)$ is a continuous dcpo, then $(P, \sigma(P))$ is compact and core compact.

Corollary 2.6. Let L be a continuous dcpo and $A (\subset L)$ a set of compact elements of L. If A is finite and $U = \uparrow A$ then the space $(U, \sigma(U))$ is compact and core compact.

Proof. The proof follows from Proposition 2.4, 2.5 and the fact that U itself is a continuous dcpo. □

Noticing that every minimal element of L is compact when L is continuous, we have also the following

Corollary 2.7. Let L be a continuous dcpo and $A (\subset L)$ a set of minimal elements of L. Then $\uparrow A$ is a Scott open set in L. If A is finite and $U = \uparrow A$, then the space $(U, \sigma(U))$ is compact and core compact.

Definition 2.1. ([5]) Let P be a dcpo. An *(internal) retraction* is a continuous map $r : P \to P$ such that $r \circ r = r$. A dcpo A is a *retract* of P if there exist Scott continuous maps $r : P \to A$ and $s : A \to P$ such that $r \circ s = id_A$. In this case, (r, s) is called a *retraction pair* and the map r is called an *(external) retraction.*

Remark 2.8. For the retraction pair (r, s), $s \circ r$ is an internal retraction on P and $s : A \to s(A)$ is an order isomorphism. If $r : P \to P$ is an internal retraction, then $r(P)$ is a retract of P. Let $p \in P$. A typical retraction is $r : P \to \downarrow p$ such that $r(x) = x$ if $x \leq p$ and $r(x) = p$ if $x \not\leq p$. It is well-known that a retract of a continuous dcpo is also a continuous dcpo.

Lemma 2.9. *([14, Theorem 4.6.2])* If L is a continuous dcpo, then L is a retract of $idl(L)$ via the retraction pair *$(sup_L(-), \twoheaddownarrow(-))$.*

Lemma 2.10. *([6, Lemma 2.12])*Let X be a topological space, and let L_1 and L be dcpo's. If L_1 is a retract of L, then $[X \to L_1]$ is a retract of $[X \to L]$.

Definition 2.2. ([11]) Let L be a dcpo. If for all $x, y \in L$ with $x \leq y$, the closed interval $[x, y]$ inheriting the order of L is a continuous dcpo, then we say that L is a *closed interval continuous* dcpo.

Lemma 2.11. *([6, Theorem 2.15])* If L is a dcpo, then the following statements are equivalent:

(1) L is a continuous L–domain;

(2) $[X \to L]$ is a continuous L–domain for all core compact spaces X;

(3) $[X \to L]$ and $[X \to J]$ are continuous dcpo's for all core compact spaces X and all closed intervals J in L;

(4) $[X \to L]$ is a continuous dcpo for all core compact spaces X;

(5) $[X \to L]$ is a continuous and closed interval continuous dcpo for all core compact spaces X.

Proof. That (1) $\Leftrightarrow$ (2) $\Leftrightarrow$ (4) has been proved in the Theorem 2.15 of [6]. It is clear that in a continuous L-domain, closed intervals are all continuous lattices. So, it follows from (1) $\Rightarrow$ (4) that (1) $\Rightarrow$ (3) $\Rightarrow$ (4) and (2) $\Rightarrow$ (5) $\Rightarrow$ (4). Thus, all the five statements are equivalent. □

3. Continuous sL-domains and their characterizations by function spaces

Definition 3.1. Let L be a dcpo. If for all $x \in L$, the principal ideal $\downarrow x$ is a join semilattice, then we say that L is an *sL-domain.*

Example 3.1. *Let $P_N = \{a, b, c_1, c_2, \ldots, c_n, \ldots\}$, where c_1 is the greatest element, a and b are incomparable and are both less then any c_i*

($i \in N$), $c_i \le c_j$ if and only if $j \le i$. Then P_N is a continuous dcpo. Because $a \vee b$ does not exist, P_N is not an sL-domain. Note that P_N can be obtained by eliminating element O_N from the dcpo M_N in [6, Fig. 5].

Example 3.2. *Let $M = \{a, b, \bar{a}, \bar{b}, 1_M\}$, where 1_M is the greatest element, a and b are incomparable, $\bar{a}$ and $\bar{b}$ are two minimal upper bounds of a and b. M can be obtained by eliminating the element O_{M_1} from the dcpo M_1 in [6, Fig. 3]. Clearly, M is a continuous dcpo, but it is not an sL-domain.*

Remark 3.1. P_N and M defined above are all continuous dcpo's. Though neither of them is an sL-domain, their closed intervals are all complete lattices. By Proposition 2.5, P_N is compact and core compact.

Proposition 3.2. Let L be a continuous sL-domain. Then L is also closed interval continuous. And for all core compact spaces X and all closed intervals J in L, $[X \to J]$ is continuous.

Proof. By Proposition 2.2, $L \cup \{\bot\}$ is a continuous L-domain. And every closed interval J of L thus is a continuous lattice. Then the proposition follows from Lemma 2.11. □

Theorem 3.3. Let X be a compact space and L a continuous dcpo. If U is a Scott open set of L, then $[X \to U]$ is a Scott open set of $[X \to L]$.

Proof. It is easy to see that $[X \to U]$ is an upper set of $[X \to L]$. Let $f_\alpha : X \to L(\alpha \in J)$ be a directed set of $[X \to L]$ and $\sup f_\alpha = f \in [X \to U]$. Then by the continuity of f_α, we have $f_\alpha^{-1}(U) \subset X$ is an open set of X ($\forall \alpha \in J$). Since for all $x \in X, \bigvee^\uparrow f_\alpha(x) = f(x) \in U$ and U is Scott open, there must be some $\alpha_0 \in J$ such that $f_{\alpha_0}(x) \in U$. So, $\{f_\alpha^{-1}(U)\}_{\alpha \in J}$ is an open cover of X. By the compactness of X, there is a finite subcover $\{f_{\alpha_i}^{-1}(U) | i = 1, 2, \ldots, n\}$. As $f_\alpha : X \to L(\alpha \in J)$ is directed, there is some $\alpha^* \in J$ such that $f_{\alpha^*} \ge f_{\alpha_i}(i = 1, 2, \ldots, n)$. So, $\forall x \in X, f_{\alpha^*}(x) \in U = \uparrow U$ and $f_{\alpha^*} \in [X \to U]$. To sum up, $[X \to U]$ is Scott open. □

Corollary 3.4. Let X be a compact space and L a continuous dcpo. If a and b are two minimal elements of L and $U = (\uparrow a) \cup (\uparrow b)$, then $[X \to U]$ is a Scott open set of $[X \to L]$.

Theorem 3.5. If L is a continuous sL–domain, then $[X \to L]$ is a continuous sL-domain for all compact and core compact spaces X.

Proof. Let $L_\bot = L \cup \{\bot\}$. Then $L_\bot$ is a continuous L-domain with L as one of its Scott open sets. By Lemma 2.11, $[X \to L_\bot]$ is a continuous L-domain. Note that since X is compact, by Theorem 3.3, $[X \to L]$ is a Scott open set of $[X \to L_\bot]$. So, $[X \to L]$ is also continuous. Since $[X \to L_\bot]$ is an L-domain, $[X \to L]$ is an sL–domain. □

Proposition 3.6. Let L be a dcpo with the least element $\bot$. If $[X \to L]$ is a continuous dcpo for all compact and core compact spaces X , then L is a continuous L-domain.

Proof. L is in fact a continuous dcpo with the least element $\bot$. Suppose that L is not an L-domain, then there exists $y \in L$ such that $\downarrow y$ is not a complete lattice. By Lemma 2.11, there is a core compact space X_0 such that $[X_0 \to \downarrow y]$ is not continuous. Construct a topological space $X_0^* = X_0 \cup \{\infty\}$ with the topology $\Omega(X_0^*) = \Omega(X_0) \cup \{X_0^*\}$. It is easy to see that X_0^* is a compact and core compact space. Thus $[X_0^* \to \downarrow y]$ is continuous by assumption and Lemma 2.10. Define $f : X_0^* \to \downarrow y$ such that $f(\infty) = \bot$, $f(X_0) = \{y\}$. It is obvious that for all $a \in \downarrow y$, $f^{-1}(\Uparrow a)$ can only be one of $\emptyset, X_0$ and X_0^*. So, f is continuous and $f \in [X_0^* \to \downarrow y]$. For every $g \in [X_0 \to \downarrow y]$, define $g^* : X_0^* \to \downarrow y$ such that $g^*(\infty) = \bot$, $g^*(t) = g(t)$ for every $t \in X_0$. It is obvious that for all $a \in \downarrow y$, $(g^*)^{-1}(\Uparrow a)$ can only be one of $\emptyset, g^{-1}(\Uparrow a)$ and X_0^*. So, g^* is continuous and $g^* \in \downarrow f$. Conversely, for every $h \in \downarrow f \subset [X_0^* \to \downarrow y]$, define $h_* = h|X_0$. Then $h_* \in [X_0 \to \downarrow y]$. It is easy to verify that for every $h \in \downarrow f$, $(h_*)^* = h$ and for every $g \in [X_0 \to \downarrow y]$, $(g^*)_* = g$. This reveals that $[X_0 \to \downarrow y]$ is order isomorphic to the principal ideal $\downarrow f$ of $[X_0^* \to \downarrow y]$. By Lemma 2.1, $[X_0 \to \downarrow y](\cong \downarrow f)$ is continuous, a contradiction! So, L is an L-domain. □

Corollary 3.7. If $P = [\bot, \top]$ is a dcpo and if for all compact and core compact spaces X, $[X \to P]$ is continuous, then P is a continuous lattice.

Corollary 3.8. Let P be a dcpo. If for all compact and core compact spaces X, $[X \to P \cup \{\bot\}]$ is a continuous dcpo, then $P \cup \{\bot\}$ is an $L-$ *domain*.

Now we have arrived at the first comprehensive theorem.

Theorem 3.9. Let L be a dcpo with a least element. Then the following statements are equivalent:

(1) L is a continuous sL–domain;

(2) L is a continuous L–domain;

(3) $[X \to L]$ is a continuous sL–domain for all compact and core compact spaces X;

(4) $[X \to L]$ is a continuous dcpo for all compact and core compact spaces X;

(5) $[X \to L]$ is a continuous and closed interval continuous dcpo for all compact and core compact spaces X.

Proof. Clearly, an sL-domain with the least element is an L-domain. The theorem follows from Lemma 2.11 and Theorem 3.5 and Proposition 3.6. □

Proposition 3.10. Let L be a continuous dcpo with a top element $\top$. If every closed interval in L is a complete lattice, then $\forall a, b \in L$ with $\twoheaddownarrow a \cap \twoheaddownarrow b \neq \emptyset$, $a \vee b$ exists.

Proof. Let $c \in \twoheaddownarrow a \cap \twoheaddownarrow b \neq \emptyset$. Let $\varphi = [c, \top]$, then $a \vee_\varphi b$ exists. We show that it is also the join of a and b in L. In fact, $a \vee_\varphi b$ is an upper bound of a and b in L. If t is another upper bound of a and b in L, then t is also an upper bound of a and b in $\varphi = [c, \top]$. So, $a \vee_\varphi b \leq t$. This shows that $a \vee_\varphi b$ is the least upper bound of a and b in L. □

Proposition 3.11. Let L be a continuous dcpo with a top element $\top$. If every closed interval in L is a complete lattice and if for any $a \in L$, we define $r_a : L \to L$ such that $\forall x \in L$,

$$r_a(x) = \begin{cases} x & \text{if } \twoheaddownarrow x \cap \twoheaddownarrow a = \emptyset \\ a \vee x & \text{if } \twoheaddownarrow x \cap \twoheaddownarrow a \neq \emptyset, \end{cases}$$

then r_a is a retraction.

Proof. By Proposition 3.10, r_a is well-defined. It is easy to see that $r_a \circ r_a = r_a$ and $\forall x \in L$, $r_a(x) \geq x$. Let $x \leq y$ in L. If $r_a(x) = x$, then $r_a(y) \geq y \geq x = r_a(x)$. If $r_a(x) = x \vee a$, then $r_a(y) = y \vee a \geq x \vee a = r_a(x)$. This shows that r_a is order preserving. Next we show that r_a is Scott continuous, i.e., r_a preserves directed joins. Let D be a directed set of L, then $r_a(D)$ is also directed. Let $k = \sup D$ and $k^* = \sup r(D)$. What is to prove is that $r_a(k) = k^*$. That $r_a(k) \geq k^*$ is clear. And clearly, $k^* = \sup r_a(D) \geq \sup D = k$. We divide the following proof into three cases:
Case 1: $k \geq a$. In this case, it is trivial that $r_a(k) = k \leq k^*$ and $r_a(k) = k^*$.
Case 2: $k \not\geq a$ and $\twoheaddownarrow k \cap \twoheaddownarrow a = \emptyset$. In this case, $k^* \leq r_a(k) = k \leq k^*$.
Case 3: $k \not\geq a$ and $\twoheaddownarrow k \cap \twoheaddownarrow a \neq \emptyset$. In this case, by the continuity of L, there is some d in D such that $\twoheaddownarrow d \cap \twoheaddownarrow a \neq \emptyset$. And thus $r_a(d) = a \vee d$. So, $k^* = \sup r_a(D) \geq a \vee \sup D = a \vee k = r_a(k) \geq k^*$ and $k^* = r_a(k)$.
To sum up, r_a is continuous and thus a retraction. □

Proposition 3.12. Let $L = \uparrow a \cup \uparrow b$ be a continuous (bicomplete) dcpo with the greatest element 1_M. If $\bar{a}$ and $\bar{b}$ are two different minimal upper bounds of a and b, then L can be retracted to $M \subset L$, where M is defined in Example 3.2.

Proof. Define $r : L \to M$ such that $\forall x \in L$,

$$r(x) = \begin{cases} a & \text{if } x \in \uparrow a \setminus \uparrow b \\ b & \text{if } x \in \uparrow b \setminus \uparrow a \\ x & \text{if } x \in \{\bar{a}, \bar{b}\} \\ 1_M & \text{otherwise,} \end{cases}$$

It is easy to verify that r is well-defined, order preserving, $r(L) = M$ and $r \circ r = r$. Next we show that r is Scott continuous, i.e., r preserves directed joins. Let D be a directed set of L, then $r(D) \subset M$ is also directed. Let $k = \sup_L D$ and $k^* = \sup_L r(D)$. What is to prove is that $r(k) = k^*$. That $r(k) \geq k^*$ is clear. Since $r(D)$ is finite and directed, $r(D)$ has its greatest element $\max r(D)$ and $k^* = \max r(D)$. We divide the following proof into several cases:
Case 1: $\max r(D) = 1_M$. In this case, it is trivial that $r(k) \leq 1_M = k^*$ and $r(k) = k^*$.
Case 2: $\max r(D) = \bar{a}$. In this case, $\bar{a} \in D$. Thus, $k \geq \bar{a}$. Since $\bar{a}$ and $\bar{b}$ are different minimal upper bounds of a and b, we have that $k \not\leq \bar{b}$. If $k \neq \bar{a}$, then since D is directed and $\sup D = k \gg a$ and $k \gg b$, there is some $d \in D$ such that $a \leq d$, $b \leq d$ and $d \not\leq \bar{a}$ and $d \not\leq \bar{b}$. For this $d \in D$, $r(d) = 1_M$, contradicting $\max r(D) = \bar{a}$! So, $k = \bar{a}$ and $r(k) = \bar{a} = k^*$.
Case 3: $\max r(D) = \bar{b}$. Similar to Case 2.
Case 4: $\max r(D) = a$. In this case, $D \subset \uparrow a \setminus \uparrow b$. Since b is a minimal element and thus a compact element, $\sup D = k \not\geq b$. This shows that $k \in \uparrow a \setminus \uparrow b$ and $r(k) = a = \max r(D) = k^*$.
Case 5: $\max r(D) = b$. Similar to Case 4.
To sum up, r preserves directed joins and r is a retraction. □

Proposition 3.13. $[P_N \to M]$ is not continuous, where P_N and M are the dcpo's defined in Examples 3.1 and 3.2.

Proof. Define $f : P_N \to M$ such that $f(a) = a, f(b) = b, f(c_i) = 1_M (i \in N)$. f preserves the order and directed joins, so is continuous. For all $n \in N$, define $g_n : P_N \to M$ such that $g_n(a) = a, g_n(b) = b, g_n(\{c_1, c_2, \ldots, c_n\}) = \{1_M\}, g_n(c_{n+1}, \ldots,) = \{\bar{a}\}$. Then g_n is continuous and $\{g_n | n \in N\}$ is directed with that $\sup g_n = f$. However, $\forall n \in N, g_n \not\geq f$, showing that $f \not\ll f$.
Suppose that $\Downarrow f$ is directed and $\sup(\Downarrow f) = f$. Then noticing the similarity in position of a and b, there must be some $h \in \Downarrow f$, $h \neq f$ and $j \in N$ such that $h(a) = a$, $h(b) = b$, $h(\{c_1, c_2, \ldots, c_j\}) = \{1_M\}, h(c_{j+1}) = \bar{b}, h(c_{j+k+1}) \leq \bar{b}$ $(\forall k \in N)$. Consider also the directed family $\{g_n | n \in N\}$

with that $\sup g_n \geq f$. It is easy to see $h \not\leq g_n (\forall n \in N)$. This shows that $h \not\ll f$, contradicting $h \in \twoheaddownarrow f$! So, $[P_N \to M]$ is not continuous. □

Theorem 3.14. Let L be a dcpo. If for all compact and core compact spaces X and all closed intervals J in L, $[X \to L]$ and $[X \to J]$ are all continuous dcpo's, then L is a continuous sL-domain.

Proof. It is obvious that L and its closed intervals are all continuous dcpo's. What is to prove is that $\forall y \in L, \downarrow y$ is a join semilattice. And this means to prove that $\forall a, b \in \downarrow y, a \vee b$ exists in $\downarrow y$.

Noticing that $\downarrow y$ is a retract of L, we have that $[X \to \downarrow y]$ and $\downarrow y$ are continuous for all compact and core compact spaces X. By the assumption, $\forall z \in \downarrow y$, $[X \to [z, y]]$ is continuous for all compact and core compact spaces X. By Corollary 3.7, $[z, y]$ is a continuous lattice. We consider the following two cases separately:

Case 1: $\twoheaddownarrow a \cap \twoheaddownarrow b \neq \emptyset$. Take $z \in \twoheaddownarrow a \cap \twoheaddownarrow b \neq \emptyset$. By Proposition 3.10, $a \vee b$ exists in $\downarrow y$.

Case 2: $\twoheaddownarrow a \cap \twoheaddownarrow b = \emptyset$. Define $r_a : \downarrow y \to \downarrow y$ such that $\forall x \in \downarrow y$,

$$r_a(x) = \begin{cases} x & \text{if } \twoheaddownarrow x \cap \twoheaddownarrow a = \emptyset \\ a \vee x & \text{if } \twoheaddownarrow x \cap \twoheaddownarrow a \neq \emptyset. \end{cases}$$

Then by Proposition 3.11, r_a is a retraction. Let $L_1 = r_a(\downarrow y)$ be the retract. It is clear that $a, b \in L_1$ and that a is a minimal element and y is the greatest element of L_1. If $a \vee_{L_1} b$ exists, then it is also the join of a and b in $\downarrow y$. So, it suffices to prove that $a \vee_{L_1} b$ exists. Clearly, L_1 with the greatest element y also fulfills the conditions in the theorem. And $\twoheaddownarrow a \cap \twoheaddownarrow b = \emptyset$ holds in L_1. Define $r_b : L_1 \to L_1$ such that $\forall x \in L_1$,

$$r_b(x) = \begin{cases} x & \text{if } \twoheaddownarrow x \cap \twoheaddownarrow b = \emptyset \\ b \vee x & \text{if } \twoheaddownarrow x \cap \twoheaddownarrow b \neq \emptyset. \end{cases}$$

Then r_b is a retraction. Let $L_2 = r_b(L_1)$ be the retract. It is clear that $a, b \in L_2$ and that a, b are minimal elements and y is the greatest element of L_2. If $a \vee_{L_2} b$ exists, then it is also the join of a and b in L_1 and in $\downarrow y$. So, it suffices to prove that $a \vee_{L_2} b$ exists. It is clear that L_2 also fulfills the conditions in the theorem and that $\twoheaddownarrow a \cap \twoheaddownarrow b = \emptyset$ holds in L_2. To prove that $a \vee_{L_2} b$ exists, let $L_3 = (\uparrow a) \cup (\uparrow b) \subset L_2$. Noticing that a, b are minimal elements in L_2, it is clear that L_3 is a Scott open set of L_2 and thus is continuous. So, by Theorem 3.3, $[X \to L_3]$ is a Scott open set of $[X \to L_2]$ and thus is continuous for all compact and core compact spaces X. By Proposition 2.5, $(L_3, \sigma(L_3))$ is a compact and core compact space. So, $[L_3 \to L_3]$ is continuous and L_3 is bicomplete. Again, if $a \vee_{L_3} b$ exists, then it is also the join of a and b in L_2 and in

$\downarrow y$. So, it suffices to prove that $a \vee_{L_3} b$ exists. As L_3 is bicomplete and has the greatest element y , the set $\mu(a,b)$ of all minimal upper bounds of a, b in L_3 is not empty. If $\mu(a,b)$ has only one element, then the element is just the join of a and b in L_3, and the theorem is proved. If $\mu(a,b)$ has more, then take two different elements $\bar{a}$ and $\bar{b}$ of $\mu(a,b)$. By Proposition 3.12, L_3 can be retract to M. And then by Lemma 2.10, $[X \to M]$ is continuous for all compact and core compact spaces X, contradicting to Proposition 3.13. So, $a \vee_{L_3} b$ exists. Thus, $a \vee b$ in $\downarrow y$ exists, as desired. □

Lemma 3.15. If L is a continuous dcpo, then $\forall w \in L$ the map $r : L \to \{w\} \cup \Uparrow w \subset L$ defined by $\forall x \in L$

$$r(x) = \begin{cases} x & \text{if } x \in \Uparrow w, \\ w & \text{otherwise.} \end{cases}$$

is a retraction.

Proof. Clearly, r is well-defined, order preserving and $r^2 = r$. What we need to prove is that r is continuous. Let D be directed in L with $\sup D = p \in L$.
Case 1 $r(p) = p \in \Uparrow w$: In this case, by the openness of $\Uparrow w$ and the directedness of D, there is some $d_0 \in \Uparrow w \cap D$. And then $r(d) = d$ for all $d \geq d_0$. Thus $\sup r(D) = p = r(\sup D) = r(p)$.
Case 2 $r(p) = w$: In this case , $\forall d \in D$, $r(d) = w$ and trivially, $\sup r(D) = w = r(\sup D) = r(p)$.
So, r preserves directed sups, r is continuous, as desired. □

Now, we can prove the second comprehensive theorem.

Theorem 3.16. Let L be a dcpo. Then the following statements are equivalent:

(1) L is a continuous sL-domain;

(2) $[X \to L]$ is a continuous sL–domain for all compact and core compact spaces X;

(3) $[X \to L]$ and $[X \to J]$ are continuous dcpo's for all compact and core compact spaces X and all closed intervals J in L.

(4) $[X \to L]$ is a continuous dcpo for all compact and core compact spaces X;

(5) $[X \to L]$ is a continuous and closed interval continuous dcpo for all compact and core compact spaces X.

Proof. (1) $\Rightarrow$ (2): Follows from Theorem 3.5.
(2) $\Rightarrow$ (1) and (2) $\Rightarrow$ (5) $\Rightarrow$ (4): Trivial.
(1) $\Rightarrow$ (3): Follows from Proposition 3.2 and Theorem 3.5.
(3) $\Rightarrow$ (1): Follows from Theorem 3.14.
(3) $\Rightarrow$ (4): Trivial.
(4) $\Rightarrow$ (3): Let $J = [z, y]$ be any closed interval in L, then $[X \to \mathop{\downarrow\!\!\!\downarrow} y]$ is continuous. By the continuity of L, $\mathop{\downarrow\!\!\!\downarrow} z \neq \emptyset$. Take a $w \in \mathop{\downarrow\!\!\!\downarrow} z$, then by Lemma 3.15, $\downarrow y$ can be retracted to $\{w\} \cup (\mathop{\uparrow\!\!\!\uparrow} w \cap \downarrow y)$. So, $[X \to \{w\} \cup (\mathop{\uparrow\!\!\!\uparrow} w \cap \downarrow y)]$ is continuous for all compact and core compact spaces X. Thus by Corollary 3.7,$\{w\} \cup (\mathop{\uparrow\!\!\!\uparrow} w \cap \downarrow y)$ is a continuous lattice. As a closed interval of this continuous lattice, $[z, y]$ itself is a continuous lattice, too. Also by Theorem 3.9, $[X \to J]$ is a continuous dcpo for all compact and core compact spaces X and all closed intervals J in L. Thus (3) is obtained. □

4. External characterizations by posets of ideals

Proposition 4.1. Let L and P be dcpo's and L an sL-domain. If $r : L \to P$ is a retraction, then P is also an sL-domain.

Proof. Let $p \in P$, we need to prove that $\downarrow p$ is a join semilattice. Let $a, b \in \downarrow p$ and $s : P \to L$ such that $rs = id_P$. Then $s(a) \vee_L s(b) \in \downarrow_L s(p)$ and $r(s(a) \vee_L s(b)) \leq r(s(p)) = p \in P$. So, $r(s(a) \vee_L s(b))$ is an upper bound of a and b in P. If t is another upper bound of a and b in P, then $s(a) \vee_L s(b) \leq s(t)$ and $r(s(a) \vee_L s(b)) \leq r(s(t)) = t$. This reveals that $r(s(a) \vee_L s(b))$ is the smallest upper bound of a and b in P. So, $\downarrow p$ is a join semilattice. □

Theorem 4.2. Let L be a continuous dcpo. Then L is a continuous sL-domain if and only if $idl(L)$ is a continuous sL-domain.

Proof. $\Rightarrow$: What we need to show is that if L is an sL-domain then $\forall K \in idl(L), \downarrow K$ is a join semilattice. Let $I, J \in \downarrow K$. Define $M = \{t \in L : (\exists a \in I, b \in J, c \in K)(a, b \in \downarrow c \text{ and } a \vee_c b \geq t)\}$. We show that $M \in idl(L)$. Clearly, M is a lower set of L. Let $s \leq a \vee_c b \in M$ and $t \leq u \vee_w v \in M$, where $a, u \in I$ with an upper bound $p \in I$ and $b, v \in J$ with an upper bound $q \in J$ and $c, w \in K$ with an upper bound $r \in K$. Then $p \vee_r q \in M$. We see $a \vee_c b = a \vee_r b$ by $r \geq c$. The same argument gives that $u \vee_w v = u \vee_r v$. Thus $s \leq a \vee_c b \leq p \vee_r q \in M$ and $t \leq u \vee_w v \leq p \vee_r q \in M$. So, M is an ideal. It is easy to see that $I, J \subset M \subset K$. Let H be another ideal with $I, J \subset H \subset K$. Then for all $a \vee_c b \in M$, where $a \in I$, $b \in J$, $c \in K$, we have $a, b \in H$ and $a, b \leq c$. Since H is an ideal, there will be an element $h \in H$, $h \geq a, b$. Then, $a \vee_h b = a \vee_c b \in H$. So, $M \subset H$. To sum up, M is the least upper

bound of I, J in $\downarrow K$, as desired.
$\Rightarrow$: Follows from Lemma 2.9 and Proposition 4.1. □

Note that for a dcpo P, $idl(P)$ is an algebraic dcpo. And if there is a dcpo D such that $P = idl(D)$, than P is an algebraic dcpo. So, we have the following theorem.

Theorem 4.3. Let L be a dcpo. Then L is a continuous sL-domain if and only if $[X \to idl(L)]$ and L are continuous dcpo's for all compact and core compact spaces X.

Proof. " $\Rightarrow$ ": Follows from Theorem 3.5 and that $idl(L)$ is an algebraic sL-domain if L is a continuous sL-domain.
" $\Leftarrow$ ": Follows from Theorem 3.16 and Theorem 4.2. □

Acknowledgement
Lemma 3.15 and Theorem 3.16(4) are due to Professor Jimmie Lawson. Thanks go to him for permitting me to publish these results in this paper. The author also would like to express thanks to the Mathematics Department of Louisiana State University for their hospitality and for providing the use of their facilities during his visit while completing this paper.

References

[1] Yixiang Chen, Stable functions and Cartesian closedness of the category of locally algebraic lattices, *Acta Mathematica Sinica*, **40**(1997), 597-602 (in Chinese).

[2] G. Gierz, et al., *A Compendium of Continuous Lattices*, Springer-Verlag, Berlin, 1980.

[3] A. Jung, *Cartesian closed categories of domains*, Ph.D. Thesis, Technischen Hochschule Darmstadt, 1988.

[4] J. D. Lawson, The duality of continuous posets, *Houston Journal of Mathematics*, **5**(1979), 357-386.

[5] J. D. Lawson, M. Mislove, Problems in domain theory and topology, in: J. Van Mill, G. M. Reed (Editors), *Open Problems in Topology*, Elsevier Sci. Publishers B.V., North-Holland, 1990, 351-372.

[6] Ying-Ming Liu, Ji-Hua Liang, Solution to two problems of J. D. Lawson and M. Mislove, *Topology and its Applications*, **69**(1996), 153-164.

[7] D. S. Scott, *Continuous lattices*, Lecture Notes in Mathematics **274**, Springer-Verlag, Berlin, 1972, 97-136.

[8] P. Taylor, An algebraic approach to stable domains, *J. Pure and Appl. Alg.* **64**(1990), 171-203.

[9] V. Stoltenberg-Hansen, I. Lindstrom, R. Griffor, *Mathematical Theory of Domains*, Cambridge Univ. Press, Cambridge, 1994.

[10] Geping Wang, Genbao Shi, The category of φ-inductive posets and a classification theorem of posets, *Chinese Annals of Mathematics*, **14**(1993), 111-117 (in Chinese).

[11] Luoshan Xu, Closed interval continuity and principal continuity on posets, *J. of Yangzhou Univ. (Nature Science Edition)*, **2(4)**(1999), 1-5 (in Chinese).

[12] Guo-Qiang Zhang, The largest Cartesian closed category of stable domains, *Theoretical Computer Science*, **166**(1996), 203-219.

[13] Dongsheng Zhao, *Generalizations of Continuous Lattices and Frames*, Doctoral Dissertation, University of Cambridge, 1992.

[14] Congyou Zheng, Lei Fan, Hongbin Cui, *Introduction to Frames and Continuous Lattices*, Capital Normal Univ. Press, Beijing, 1994 (in Chinese).

Chapter 7

PROJECTIVES AND INJECTIVES IN THE CATEGORY OF QUANTALES*

Yong-Ming Li
Department of Mathematics
Shaanxi Normal University
Xi'an 710062, P.R. China
liyongm@snnu.edu.cn

Meng Zhou
Department of Applied Mathematics
Beijing University of Aeronautics and Astronautics
Beijing 100083, P.R.China

Abstract The notions of multiplication-stable completely distributive lattices and weakly multiplication-stable completely distributive lattices are introduced in this paper. It is proved that the regular projective quantales are weakly multiplication-stable completely distributive lattices and multiplication-stable completely distributive lattices are regular projective quantales. For the class $\mathcal{E}$ of all onto quantale homomorphisms whose right adjoints preserve multiplication operations, it is proved that $\mathcal{E}$-projective quantales are exactly the weakly multiplication-stable completely distributive lattices. Finally, it is also proved that there are no nontrivial injective objects in the category of quantales.

Keywords: Quantale, projective quantale, injective quantale,, multiplication-stable completely distributive lattice, supercoherent quantale.

*This work was supported by National Science Foundation of China (Grant No: 60174016,10226023), "TRAPOYT" of China and the Key Project of Fundamental Research (Grant No: 2002GB312200).

GQ Zhang et al. (eds.), Domain Theory, Logic and Computation, 151-160.

1. Introduction

Quantales were introduced by Mulvey in order to provide a lattice theoretic setting for studying non-commutative C*-algebras, as well as constructive foundations for quantum mechanics (see [12, 13]). The study of such partially ordered algebraic structures goes back to a series of papers by Ward and Dilworth (see [5, 18, 19]) in the 1930s. They were motivated by the ideal theory of commutative rings. Following Mulvey, various types and aspects of quantales have been considered by many researchers, see the reference [16] by Rosenthal.

Even more recently, quantales have been arisen in analyzing in the semantics of linear logic, a logic system developed by Girard [7], which supports part of the foundation of theoretical computer science, see [11, 20]. Quantales also have been linked to the operational semantics and denotational semantics of computer languages, they give the process semantics of various kinds of observational equivalence, see [1, 10].

So it is important to study the lattice properties of quantales. In this note, we study the projectives and injectives of quantales as a category.

Let $\mathcal{C}$ be a category and E a set of epimorphisms of $\mathcal{C}$, then an object c of $\mathcal{C}$ is said to be projective w.r.t E (simply E-projective) if for any morphism $f : c \to b$ and a morphism $g : a \to b$ in E, there is a morphism $\overline{f} : c \to a$ such that $f = g \circ \overline{f}$. Dually, one defines an M-injective in a category $\mathcal{C}$ for M a set of monomorphisms of $\mathcal{C}$. Ususally, we consider regular projectives and regular injectives in a category $\mathcal{C}$.

2. Regular projectives in the category of quantales

A quantale is a complete lattice Q with an associative binary multiplication $*$ satisfying

$x * (\bigvee_{i\in I} x_i) = \bigvee_{i\in I}(x * x_i)$ and $(\bigvee_{i\in I} x_i) * x = \bigvee_{i\in I}(x_i * x)$

for all $x, x_i \in Q$, $i \in I$, (I is a set). 1 denotes the greatest element of Q, 0 is the smallest element of Q. A quantale Q is said to be unital if there is an element $u \in Q$ such that $u * a = a * u = a$ for all $a \in Q$.

Let P and Q be quantales, a function $f : P \to Q$ is a homomorphism of quantales if f preserves arbitrary sups and it also preserves the operation $*$. If P and Q are unital, then f is a unital homomorphism if in addition to being a homomorphism, it satisfies $f(u_P) = u_Q$, where u_P and u_Q are the respective units of P and Q.

Let **Quant** denote the category of quantales and homomorphisms and **UnQuant** be the category of unital quantales and unital homomorphisms. Then a regular epimorphism in **Quant** is just an onto homomorphism, and a monomorphism in **Quant** is just a one-to-one (or embedding) homomorphism. In this case, a quantale S is a regular projective in **Quant** if for any homomorphism $f : S \to Q$ and an onto homomorphism $g : P \to Q$, there is a homomorphism $\overline{f} : S \to P$ such that $g \circ \overline{f} = f$. Dually, an injective in **Quant** can be defined. For facts concerning quantales in general we refer to [16].

In any complete lattice L, let $a \lhd b$ (read as "a is wedge below b") mean that, for any subset A of L, $b \leq \bigvee A$ implies that $a \leq t$ for some $t \in A$. Then it is well known that (see [6]) L is a completely distributive lattice (short by CDL in this paper) iff $x = \bigvee\{y|y \lhd x\}$ for all $x \in L$.

Definition 1. A quantale $(L, *)$ is called a $*$-stable CDL if it satisfies the following conditions:

1)L is a CDL;

2)$x_1 \lhd y_1, x_2 \lhd y_2 \Rightarrow x_1 * x_2 \lhd y_1 * y_2$;

3)$x \lhd y_1 * y_2 \Rightarrow \exists x_1 \lhd y_1, \exists x_2 \lhd y_2$, such that $x = x_1 * x_2$.

A quantale $(L, *)$ is called a weakly $*$-stable CDL if it satisfies the above conditions 1), 2) and the following conditions:

4)$x \lhd y_1 * y_2 \Rightarrow \exists x_1 \lhd y_1, \exists x_2 \lhd y_2$, such that $x \leq x_1 * x_2$.

Obviously, a $*$-stable CDL is also a weakly $*$-stable CDL.

For any set X, the free quantale generated by X is just powerset $\mathcal{P}(X^*)$ under subset inclusion, where X^* is the set of all finite strings of X, which is the free semigroup generated by X with the juxtaposition multiplication. The multiplication on $\mathcal{P}(X^*)$ is given by, for $A, B \in \mathcal{P}(X^*)$, $A * B = \{a * b | a \in A, b \in B\}$. Let $X^*{}_0 = X^* \cup \{\text{the empty string}\}$, then $X^*{}_0$ is free monoid generated by X, and $\mathcal{P}(X^*{}_0)$ is free unital quantale generated by X.

For any quantale $(S, *)$, the free quantale generated by the set S is $\mathcal{P}(S^*)$, define the map $\bigvee_S : \mathcal{P}(S^*) \to S$ by

$\bigvee_S(A) = \bigvee_S\{x_1 * ... * x_n \in S | x_1 ... x_n \in S^*\}$

then $\bigvee_S$ is obvious a quantale homomorphism. In the following, if there is no confusion, we do not distinguish $x_1...x_n \in S^*$and $x_1 * ... * x_n \in S$.

Proposition 1. *A quantale S is regular projective iff there exists a quantale homomorphism $h : S \to \mathcal{P}(S^*)$ such that $\bigvee_S \circ h = id_S$.*

Proof. Suppose that S is projective. The homomorphism $\bigvee_S$ is a regular epimorphism, so by the definition of projectives, for the identity map $id_S : S \to S$, there is a homomorphism $h : S \to \mathcal{P}(S^*)$ such that $\bigvee_S \circ h = id_S$.

Conversely, let $h : S \to \mathcal{P}(S^*)$ be a homomorphism satisfying $id_S = \bigvee_m \circ h$. Given any onto homomorphism $f : P \to Q$ and a morphism $g : \mathcal{P}(S^*) \to Q$. For each $s \in S$, there exists a $x_s \in P$ such that $g(\{s\}) = f(x_s)$. Then we obtain a map $k : S \to P$ by $k(s) = x_s$. Now $\mathcal{P}(S^*)$ is freely generated by the set S, there is a homomorphism $\overline{k} : \mathcal{P}(S^*) \to P$ such that $\overline{k} \circ \{\}_S = k$, where $\{\}_S : S \to \mathcal{P}(S^*)$ is the natural inclusion map. From S generates $\mathcal{P}(S^*)$ and $g \circ \{\}_S = f \circ k$, it follows that $g = f \circ \overline{k}$. This shows that $\mathcal{P}(S^*)$ is projective. Now let $f : P \to Q$ be any onto homomorphism and $g : S \to Q$ be any homomorphism. Then $g \circ \bigvee_S : \mathcal{P}(S^*) \to Q$ is a homomorphism from $\mathcal{P}(S^*)$ to Q, $\mathcal{P}(S^*)$ is projective, there is a $j : \mathcal{P}(S^*) \to P$ such that $g \circ \bigvee_S = f \circ g$. Then the homomorphism $k = j \circ h : S \to P$ satisfies $f \circ k = (f \circ j) \circ h = g \circ \bigvee_S \circ h = g$. Hence S is projective. □

Lemma 1. *If $h : L \to M$, $g : M \to L$ both are quantale homomorphisms and $g \circ h = id_L$, and M is a weakly $*$-stable CDL, then L is also a weakly $*$-stable CDL.*

Proof. First, we have the following implication

$y \lhd h(x)$ in $M \Rightarrow g(y) \lhd x$ in L. (*)

This is because, if $x \leq \bigvee_L A$ in L , then $h(x) \leq h(\bigvee_L A) = \bigvee_M h(A)$ in M. Since $y \lhd h(x)$ in M , $y \leq h(a)$ for some $a \in A$, and then $g(y) \leq gh(a) = a$ for some $a \in A$, hence $g(y) \lhd x$ in L. Since M is CDL , $h(b) = \bigvee\{y \in M | y \lhd h(b)\}$ for all $b \in L$, and then $b = gh(b) = \bigvee\{g(y) \in L | y \lhd h(b)\}$. From (*)it follows that $b = \bigvee\{l \in L | l \lhd b\}$, thus L is a CDL.

Furthermore, if $a_1 \lhd b_1, a_2 \lhd b_2$ in L, from $b_i = \bigvee\{g(y_i) | y_i \lhd h(b_i)\}$, it follows that $a_i \leq g(y_i)$ for some $y_i \lhd h(b_i)$, then $a_1 * a_2 \leq g(y_1) * g(y_2) = g(y_1 * y_2)$, where $y_1 * y_2 \lhd h(b_1) * h(b_2) = h(b_1 * b_2)$. By the hypothesis of g, M and h, the latter implies $g(y_1 * y_2) \lhd b_1 * b_2$ in L, and hence we have $a_1 * a_2 \lhd b_1 * b_2$.

Finally, if $a \lhd b_1 * b_2$ in L, since $b_1 * b_2 = \bigvee\{g(y) | y \lhd h(b_1 * b_2)\}$, there is $y \lhd h(b_1) * h(b_2)$ such that $a \leq g(y)$. Since M is $*$-stable CDL and $y \lhd h(b_1) * h(b_2)$ in M, there is $y_1 \lhd h(b_1), y_2 \lhd h(b_2)$ such that $y \leq y_1 * y_2$, then from (*), it follows that $a_1 = g(y_1) \lhd b_1, a_2 = g(y_2) \lhd b_2$ and $a \leq g(y) \leq g(y_1) * g(y_2) = a_1 * a_2$.

Hence L is also a weakly $*$-stable CDL. □

Lemma 2.*The free quantale $\mathcal{P}(X^*)$ generated by X is a $*$-stable CDL.*

Proof. 1) $\mathcal{P}(X^*)$ is a power set of X^*, and hence a CDL.

2)$x \lhd A$ in $\mathcal{P}(X^*)$ iff $x \in A$ for some $x \in X^*$, hence if $x_1 \lhd A_1, x_2 \lhd A_2 \Rightarrow x_1 \in A_1, x_2 \in A_2 \Rightarrow x_2 * x_2 \in A_1 * A_2 \Rightarrow x_1 * x_2 \in A_1 * A_2$.

3)If $x \lhd A_1 * A_2$ then $x \in A_1 * A_2 = \{x_1 * x_2 | x_1 \in A_1, x_2 \in A_2\}$, and hence $x = x_1 * x_2$ for some $x_1 \in A_1, x_2 \in A_2$, that is, $x = x_1 * x_2$ for some $x_1 \in A_1, x_2 \in A_2$.

Therefore, $\mathcal{P}(X^*)$ is a $*$-stable CDL. □

Theorem 1. *If S is a regular projective quantale , then S is a weakly $*$-stable CDL; and if S is a $*$-stable CDL, then S is a regular projective quantale.*

Proof. If S is projective, then from Lemma 1 and Lemma 2, S is a weakly $*$-stable CDL.

Suppose that S is a $*$-stable CDL, if we can prove that $\bigvee_S : \mathcal{P}(S^*) \to S$ has a left inverse $h : S \to \mathcal{P}(S^*)$, i.e. $\bigvee_S \circ h = id_S$, then, by Proposition 1, S is regular projective quantale.

Define $h : S \to \mathcal{P}(S^*)$as follows:

$h(s) = \{x \in S^* | x \lhd s \text{ in } S\}$.

We check that h is a quantale homomorphism in the following.

Since $h(\bigvee s_i) = \{x \in S^* | x \lhd \bigvee s_i \text{ in } S\} = \{x \in S^* | x \lhd s_i \text{ for some } i \text{ in } S\} = \bigcup\{x \in S^* | x \lhd s_i \text{ in } S\} = \bigcup h(s_i)$, it follows that $h(\bigvee s_i) = \bigvee h(s_i)$.

Since $h(x_1 * s_2) = \{x \in S^* | x \lhd s_1 * s_2 \text{ in } S\} = \{x \in S^* | \exists x_1 \lhd s_1, x_2 \lhd s_2,$ such that $x = x_1 * x_2$ in $S\}$, and $h(s_1) * h(s_2) = \{x_1 * x_2 \in S^* | x_1 \lhd s_1, x_2 \lhd s_2\}$, it follows that $h(s_1 * s_2) = h(s_1) * h(s_2)$.

Furthermore, since S is a CDL, $\bigvee_S \circ h(s) = \bigvee_S\{x \in S | x \lhd s\} = s$ for any $s \in S$, $\bigvee_S \circ h = id_S$. □

We further give a characterization of a weakly $*$-stable CDL in the following.

Let A be a partially ordered set with an associate binary multiplication satisfying

$x \leq y \Rightarrow x * z \leq y * z$ and $z * x \leq z * y$

for all $x, y, z \in A$. Then A is called a partially ordered semigroup (short by po-semigroup in this paper). For a po-semigroup A, let DA denote the down-set lattice of partially ordered set A, that is, the lattice of all $X \subseteq A$ such that $a \in X$ implies $b \in X$ for all $b \leq a$. DA is closed under arbitrary unions and intersections, hence a complete lattice, and $X \lhd Y$ in DA iff $X \subseteq \downarrow a$ for some $a \in Y$. Multiplication on DA is defined as follows

$X * Y = \downarrow \{x * y | x \in X, y \in Y\} = \{z \in A | \exists x \in X, \exists y \in Y$ such that $z \leq x * y\}$

for all $X, Y \in DA$. Then it is an easy task to check that under the above multiplication DA become a quantale. Concerning this quantale, we have the following proposition to characterize a weakly $*$-stable CDL.

Lemma 3. *For any po-semigroup A, DA defined as above is a weakly $*$-stable quantale.*

Proof. First, since DA is a sub-complete lattice of powerset lattice $\mathcal{P}(A)$, DA is a CDL.

Second, if $X_1 \lhd Y_1$ and $X_2 \lhd Y_2$ in DA, then $X_1 \subseteq \downarrow a_1$ and $X_2 \subseteq \downarrow a_2$ for some $a_1 \in Y_1, a_2 \in Y_2$, it follows that $X_1 * X_2 \subseteq \downarrow a_1 * \downarrow a_2 = \downarrow (a_1 * a_2)$ and $a_1 * a_2 \in Y_1 * Y_2$ and thus $X_1 * X_2 \lhd Y_1 * Y_2$.

Third, if $X \lhd Y_1 * Y_2$, then $X \subseteq \downarrow (a_1 * a_2) = \downarrow a_1 * \downarrow a_2$ for some $a_1 \in Y_1, a_2 \in Y_2$. Let $X_1 = \downarrow a_1, X_2 = \downarrow a_2$, then $X \leq X_1 * X_2$ and $X_1 \subseteq \downarrow a_1, X_2 \subseteq \downarrow a_2$ for some $a_1 \in Y_1, a_2 \in Y_2$, the latter implies that $X_1 \lhd Y_1$ and $X_2 \lhd Y_2$.

Hence, DA is a weakly $*$-stable quantale. □

If S is a quantale, then S is also a po-semigroup. In this case, we have an onto map $\bigvee_S : DS \to S$ defined as follows:

$\bigvee_S(X) = \bigvee_S X$

for any $X \in DS$. It is readily verified that $\bigvee_S$ is a homomorphism.

Proposition 2. *A quantale S is a weakly $*$-stable quantale iff there is a right inverse of $\bigvee_S$, that is, there is a homomorphism $h : S \to DS$ such that $\bigvee_S \circ h = id_S$.*

The proof is similar to that of Theorem 1, we omit it here.

Furthermore, weakly $*$-stable quantales are just certain projective objects in the category of quantales which we describe in the following.

For two po-semigroups A and B, a function $f : A \to B$ is called a po-semigroup homomorphism if f preserves multiplications and orders. Let **Pg** denote the category of all po-semigroups with po-semigroups homomorphisms, then obviously, **Quant** is a (non-full) subcategory of **Pg**, and the corresponding $A \to DA$ defines a functor $D : Pg \to Quant$ such that, for any $f : A \to B$ in **Pg**, $Df : DA \to DB$ take each $X \in DA$ to $\bigcup \downarrow f(a)(a \in X)$, that is, the downset in B generated by $f(X)$. Concerning the functor, and letting $\mathcal{E}$ denote all the onto quantale homomorphisms whose right adjoints preserve multiplication, we have the following two lemmas.

Lemma 4. *D is left adjoint to the inclusion functor $Quant \to Pg$.*

Proof. It suffices to show that $\downarrow: A \to DA$ is universal among the po-semigroup homomorphisms from A into quantales for a quantale A. Assuming that $f : A \to B$ is a po-semigroup homomorphism for a quantale B, define $\overline{f} : DA \to B$ as $\overline{f}(D) = \bigvee_B\{f(d)|d \in D\}$ for and $D \in DA$. Then it is readily verified that $\overline{f}$ is a unique quantale homomorphism satisfying $\overline{f} \circ \downarrow = f$. □

Lemma 5. *Each DA, $A \in Pg$, is $\mathcal{E}$-projective in the category of quantales.*

Proof. Consider the diagram

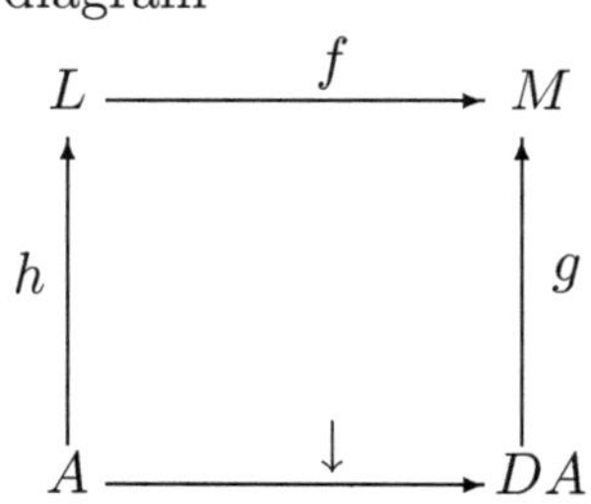

where $h = f^{\vee} \circ g \circ \downarrow$, $f^{\vee}$ the right adjoint of f, given by $f^{\vee}(z) = \bigvee\{x|f(x) \leq z\}$. Let $k : DA \to L$ be the unique quantale homomorphism such that $k \circ \downarrow = h$, resulting from Lemma 4 and the fact that h is a po-semigroup homomorphism because $f^\vee$ is. Now, $f \circ f^\vee = id_M$ whenever f is surjective, and then

$f \circ k \circ \downarrow = f \circ h = f \circ f^\vee \circ g \circ \downarrow = g \circ \downarrow$,

which shows $g = f \circ k$. Hence DA is $\mathcal{E}$-projective. □

Combining the above lemma and Proposition 2, we can obtain the following:

Theorem 2. *The $\mathcal{E}$-projectives in* **Quant** *are exactly the weakly $*$-stable CDLs.*

Remark 1: Using the same techniques, we can give the same results for the regular projective objects and $\mathcal{E}$-projective objects in the category of **UnQuant**. We only add extra condition that $u \lhd u$ for unit of quantale for the definition of (weakly) $*$-stable CDL.

Problem. *Is there a weakly $*$-stable CDL which is not a $*$-stable CDL?*

Examples:

(1)Any complete lattice L can be made into a quantale by taking $x * y = y$ for all $x, y \in L$ with $y \neq 0$ and $x * 0 = 0$. Then $*$-quantales are exactly complete lattices with sup-preserving mappings, the latter category is denoted by **Sup**. In this case,

weakly $*$-stable CDLs=$*$-stable CDLs=CDLs,

hence, the regular projective objects in **Sup** are CDLs, which was shown in [4] by Crown.

(2)If the meet operation distributes over the multiplication operation, that is,

$a \wedge (b * c) = (a \wedge b) * (a \wedge c)$,

then it is easily verified that the notion of a weakly $*$-stable CDL is the same as that of a $*$-stable CDL. In these cases, the regular projective quantales are exactly weakly $*$-stable CDL. In particular, if $* = \wedge$, finite meets, then $*$-quantales are just frames with frame homomorphisms (for general theory of frames, see [9]). In this case, the weakly $*$-stable CDLs are exactly stably supercontinuous lattices defined by Banaschewski in [2,3], and they are exactly the regular projectives in the category of frames.

(3)The unit interval $[0, 1]$ with usual order is a complete chain and thus a CDL. If $*$ is an associative and commutative binary multiplication which is also non-decreasing about its first variable, that is, $x \leq y \Rightarrow x * z \leq y * z$, and $x * 1 = x$ for any $x \in [0, 1]$, then $*$ is called a triangular norm or t-norm (see [17]). If $*$ is a continuous (w.r.t. the usual topology of $[0, 1]$) triangular norm (e.g., product or minimum operation), then $([0, 1], *)$ is a quantale.

Product operation is a continuous t-norm on $[0, 1]$, we check that $([0, 1], *)$ is a $\cdot$-stable CDL in the following. Note $x \lhd y$ in $[0, 1]$ iff $x < y$

in $[0,1]$. If $x_1 < y_1$ and $x_2 < y_2$ in $[0,1]$, then $x_1x_2 < x_1y_2 < y_1y_2$. And if $x < y_1y_2$, choose $x_1 < y_1$ such that $x < x_1y_2$ and let $x_2 = x/x_1$, then $x_2 < y_2$ and $x = x_1x_2$. Therefore, $([0,1], \cdot)$ is a $\cdot$-stable CDL.

In fact, if $*$ is any continuous t-norm on $[0,1]$ such that the mapping $a * x : [0,1] \to [0,1]$ is strictly increasing for any fixed $a \in (0,1)$, then $([0,1], *)$ is a $*$-stable CDL. Since $1 \lhd 1$ is not true in $[0,1]$, these examples are not regular projective in the category of unital quantales.

3. Injective objects in the category of quantales

Lemma 6. *Every quantale can be embeded in a unital quantale.*

Proof. Given a quantale S, define $\overline{S} = S \times \{0,1\}$as a complete lattice with the cartesian product of S and $\{0,1\}$, where $0,1 \in S$. Define the multiplication on $\overline{S}$ as follows,

$(r_1, k_1) * (r_2, k_2) = (r_1 * r_2 \vee (r_1 \wedge k_2) \vee (r_2 \wedge k_1), k_1 \wedge k_2)$.

It is easily task to check that $\overline{S}$ is a unital quantale with the unit $(0,1)$. The embedding $m : S \to \overline{S}$ is defined as follows

$m(r) = (r, 0)$

for all $r \in S$. □

Note that the above construction corresponds to the extension of a non-unital ring to a unital ring (see [8]). In fact, the above construction $\overline{S}$ is quantale isomorphic to the construction $K[e]$ given in [14].

Lemma 7. *If $e : Q \to S$ is a onto quantale homomorphism and Q has an unit u, then S has unit $e(u)$.*

It is an easy task to check.

Corollary 1. *Injective quantale must be unital.*

Proof. Combining Lemma 6 and Lemma 7 deduce the result. □

For any complete lattice L, let $Sup(L)$ denote the unital quantale of sup-preserving maps from L to itself, with the supremum given by the pointwise ordering of mappings, with the multiplication corresponding to the composition of mappings, and with the unit given by the identity mapping. Then $Sup(L)$ is a simple quantale, as studied in [14,15], that is, any surjective homomorphism of quantale from it is either an isomorphism or a constant homomorphism.

Theorem 3. *There are no nontrivial injective quantales.*

Proof. For any unital quantale S, the map $m : S \to Sup(S)$ defined as follows

$m(a)(b) = a * b$

is a quantale embedding (for the detailed proof, see [14]). First, we prove that m is not an onto map if S contains more than two elements. For any $a \in S$, define a map $f_a : S \to S$ as $f_a(b) = a$ if $b \neq 0$ and $f_a(0) = 0$, then f_a is a sup-preserving map, i.e., $f_a \in Sup(S)$. If m is an onto map, then, for any $a \in S$, there exists an $a_1 \in S$ such that $m(a_1) = f_a$. Since S is a unital quantale and S contains more than two elements, the unit u is not zero. We have $a_1 = a_1 * u = m(a_1)(u) = f_a(u) = a$. Hence $m(a) = f_a$ for any $a \in S$. But the identity map $id_S : S \to S$ in $Sup(S)$ can not write the form f_a for any $a \in S$ if S contains more than two elements, so id_S is not any image of m. This shows that m is not an onto map, which contradicts with the assumption that m is an onto map. Therefore, m is not an onto map if S contains more than two elements.

Second, if S is an injective quantale, then S is unital and the map m is a quantale embedding. From the definition of injective quantale, for the identity map $id_S : S \to S$, there is an onto homomorphism $e : Sup(S) \to S$ such that $e \circ m = id_S$. Since $Sup(S)$ is a simple quantale, e is an isomophism or a constant homomorphism. If e is a constant homomorphism, then S is a singleton quantale. If e is an isomorphism, then m is also an isomorphism. We have shown that m is not an onto map if S contains more than two elements, hence S should be a singleton quantale or a two-point frame $\{0, 1\}$. If S is the two-point frame $\{0, 1\}$, then S is not injective in the category of frames ([3]), and thus S is not injective in the category of quantales. Therefore, an injective must be a singleton, there are no nontrivial injective quantales. □

References

[1] S. Abramsky, S. Vickers, Quantales, observational logic, and process semantics, Math. Struct. Comp. Sci., 3(3) (1993), 161-227

[2] B. Banaschewiski, On the topologies of injective spaces, in: Continuous Lattices and Their Applications, Lecture Notes in Pure and Applied Mathematics 101 (Marcel Deckker, New York, 1985) 1-8

[3] B. Banaschewiski,S.B. Niefield, Projective and supercoherent frames, J. Pure Appl. Alg., 70 (1991), 45-51

[4] G.W. Crown, Projectives and injectives in the category of complete lattices with residuated mappings, Math. Ann., 187 (1970), 295-299

[5] R.P. Dilworth, Noncommutative residuated lattices, Tanns. Amer.Math.Soc., 46 (1939), 426-444

[6] G. Gierz,etc., A Compendium of Continuous Lattices, (Springer-Verlag, Birlin,1980).

[7] J.Y. Girard, Linear logic, Theor. Comp. Sci.,50 (1987), 1-102

[8] T.W. Hungerford, Algebra, (Springer-verlag, Berlin 1980)

[9] P.T. Johnstone, Stone spaces, (Cambridge University Press, Cambridge, 1982)

[10] Y.M. Li, Z.H. Li, Quantales and process semantics of bisimulation, Acta Math. Sinica, 42 (3) (1999), 313-320 (in Chinese)

[11] Y.M. Li, Non-commutative linear logic and its quantale semantics, J. Shaanxi Normal University, 29 (2) (2001),1-5 (in Chinese)

[12] C.J. Mulvey, &, Suppl.Rend.Circ.Mat.Palermo Ser.II,12 (1986), 99-104

[13] C.J. Mulvey, J.W. Pelletier, On the quantisation of point, J. Pure Appl. Alg.,159 (2001), 231-295

[14] J. Paseka, D. Kruml, Embeddings of quantales into simple quantales, J. Pure Appl. Alg.,148 (2000), 209-216

[15] J.W. Pelletier, J. Rosicky, Simple involutive quantales, J.Algebra, 195 (1997), 367-386

[16] K.I. Rosenthal, Quantales and their application, (Longman Scientific and Technical, London, 1990)

[17] B. Schweizer, A. Sklar, Statistical metric spaces, Pacific J. Math., 10(1960), 313-334

[18] M. Word, Structure residuation, Ann. Math., 39 (1938), 558-569.

[19] M. Ward, R.P. Dilworth, Residuated lattices, Trans. Amer. Math. Soc., 45 (1939), 335-354

[20] D. Yetter, Quantales and noncommutative linear logic, J.Symbolic Logic, 55 (1990), 41-64

Chapter 8

ON MINIMAL EVENT AND CONCRETE DATA STRUCTURES

Felipe Bracho
IIMAS, Universidad Nacional Autónoma de México
01000 México, D.F., México
bracho@servidor.unam.mx

Manfred Droste and Ingmar Meinecke*
Institut für Algebra, Technische Universität Dresden
D-01062 Dresden, Germany
{droste,meinecke}@math.tu-dresden.de

Abstract We consider concrete domains and event domains as well as their generating structures, *i.e.*, concrete data structures and event structures. Using certain equivalence relations on the prime intervals of the domains we can associate different event structures and cds to a given domain, respectively. By introducing reductions we show the existence of a greatest event structure and cds as well as of minimal event structures and cds generating a given event and concrete domain, respectively.

Keywords: Event structure, concrete data structure, event domain, concrete domain, reduction, minimization

1. Introduction

In the theory of denotational semantics of programming languages, various kinds of systems of information and associated partial orders (domains) of information have been studied. Scott [Sco82] considered information systems and domains. Concrete data structures and concrete

*Supported by *Deutsche Forschungsgemeinschaft* (German Research Foundation)

GQ Zhang et al. (eds.), Domain Theory, Logic and Computation, 161-178.

domains were introduced by Kahn and Plotkin [KP78, KP93] in order to allow a fairly general semantical definition of sequentiality, see also Berry [Ber78] and Berry and Curien [BC82]. Winskel [Win81, Win87] investigated a generalization, the event structures and event domains. Event structures also bear a close relationship (categorical adjunctions) with models for concurrency like Petri nets, Mazurkiewicz traces or automata with concurrency relations, see [NPW81], [DS02] and [WN95] for a survey. A variety of further results can be found in [Cur93].

In this paper we show that given an event domain $(D, \leq)$, with respect to the present notion of reductions there always exists a greatest trim event structure $\mathcal{E}_{max}$ generating $(D, \leq)$, and any trim event structure generating $(D, \leq)$ can be reduced to a minimal one $\mathcal{E}_{min}$ also generating $(D, \leq)$. The analogous results are proved for concrete domains and trim cds.

An *event structure* $\mathcal{E}$ consists of a set E of events together with a consistency predicate for finite subsets of E and an enabling relation between consistent subsets and elements of E satisfying certain natural axioms. In the more restricted notion of a *concrete data structure* or *cds* $\mathcal{M}$ each event is linked to a cell, which may be filled by a value. A *state* or *configuration* of events is a subset X of E such that each finite subset of X is consistent, and each element of X can be deduced, starting from the empty set, through finitely many successive applications of the enabling relation for finite subsets and new elements of X. The set of all such states of $\mathcal{E}$ and $\mathcal{M}$, respectively, partially ordered by inclusion, is denoted by $(D(\mathcal{E}), \subseteq)$ and $(D(\mathcal{M}), \subseteq)$. Droste [Dro89] gave with the notion of an *event domain* an order-theoretic characterization of all the partial orders $(D(\mathcal{E}), \subseteq)$ where $\mathcal{E}$ is an event structure. This generalizes results of Winskel [Win81, Win87] who obtained the corresponding representation theorems under the additional assumption that either $\mathcal{E}$ is stable or that the consistency relation is induced by a binary relation of conflict on E. The characterization of the partial orders $(D(\mathcal{M}), \subseteq)$, where $\mathcal{M}$ is a cds, was given by Kahn and Plotkin [KP78]. They denoted these partial orders as *concrete domains*.

As is well-known, a given event or concrete domain may be generated by many non-isomorphic structures. This indicates that we should endow the domains with more structure in order to undo the loss of information. Indeed, as we will show here, $\mathcal{E}$ and $\mathcal{M}$ naturally induce an equivalence relation on the prime intervals of $(D(\mathcal{E}), \subseteq)$ and $(D(\mathcal{M}), \subseteq)$. Using these equivalence relations, called *congruences*, we give constructions for associating event structures or cds to a given event or concrete domain, respectively. These associated structures do not contain superfluous events; therefore we call such structures *trim*.

For a given event or concrete domain $(D, \leq)$ the class of all trim event structures $\mathcal{E}$ or cds $\mathcal{M}$ generating $(D, \leq)$ respectively, can be structured by the notion of a *reduction* between two structures. A reduction is defined to be an epimorphism preserving the consistency and enabling relation. Up to isomorphism such a reduction does not change the induced domains. Recall that in classical formal language theory, it is known that each automaton accepting a given language can be reduced to a minimal one recognizing the same language. In this paper, we show that there always exists a greatest trim event structure $\mathcal{E}_{max}$ and minimal trim structures $\mathcal{E}_{min}$ generating the given event domain $(D, \leq)$. The same holds true for trim cds generating a given concrete domain $(D, \leq)$. However, we present examples to show that in general neither $\mathcal{E}_{min}$ nor $\mathcal{M}_{min}$ are unique up to isomorphism.

Similar questions were considered for other structures as well. Bracho and Droste [BD94] considered deterministic automata with concurrency relations and their induced weak concurrency domains of concurrent computation sequences. They obtained a greatest automaton and minimal automata generating the same domain. Moreover, if the set of events of the automaton is fixed, there is even a least structure. Similar results were obtained by Kuske [Kus94] for nondeterministic automata with concurrency relations. An approach towards uniqueness of event structures generating a given event domain was given in [Dro89]. Most results of this paper were obtained by Bracho and Droste [BD97] for event structures and by Meinecke [Mei00] for concrete data structures and their domains, respectively. A full version of this paper with additional categorical results between labeled event domains and event structures, respectively between labeled concrete domains and cds, can be found in [BDM01].

2. Event structures, concrete data structures and their domains

In this section, we recall for the convenience of the reader the necessary background from concrete data structures, event structures and event and concrete domains as needed in the following. For more information on this, we refer the reader to [KP78, KP93, Win81, Win87] and the monograph [Cur93]. For any set E let $Fin(E)$ comprise all finite subsets of E.

Definition 2.1 (*cf.* [KP93, Def. 6.16], [Cur93, Def. 2.1.1]).
A *concrete data structure* or, for short, a *cds* $\mathcal{M} = (C, V, E, \vdash)$ is given by three countable sets C, V and E of *cells*, *values* and *events* and the *enabling relation* $\vdash$ such that:

(1) $E \subseteq C \times V$ and $\forall c \in C\, \exists v \in V : (c, v) \in E$,

(2) $\vdash \subseteq Fin(E) \times C$.

We say that $\{e_0, \ldots, e_n\}$ is an *enabling* of c if $\{e_0, \ldots, e_n\} \vdash c$, which we write simply $e_0, \ldots, e_n \vdash c$. If a cell c is enabled by the empty set we call c *initial* and denote this by $\vdash c$. A subset $X \subseteq E$ is a *state* of $\mathcal{M}$ if the following conditions are satisfied:

(1) *consistency*: if $(c, v_1), (c, v_2) \in X$, then $v_1 = v_2$,

(2) *safety*: for any $e = (c, v) \in X$ there exists a sequence of events $e_0, \ldots, e_n$ such that $e_n = e$, $e_i = (c_i, v_i) \in X$ and $\{e_j : 0 \leq j < i\}$ contains an enabling of c_i for all $i \leq n$.

A sequence as described in the second condition is called a *deduction*. $Cons(E)$ denotes the set of all finite and consistent subsets of E. The set of all states of a cds $\mathcal{M}$, ordered by inclusion, is denoted by $(D(\mathcal{M}), \subseteq)$ and called the *concrete domain associated with* $\mathcal{M}$. If a poset $(D, \leq)$ is isomorphic to $(D(\mathcal{M}), \subseteq)$ we say that the cds $\mathcal{M}$ generates $(D, \leq)$. We shall see later that $(D(\mathcal{M}), \subseteq)$ has rich order-theoretic properties.

Within the notion of a cds causal dependence of events is given by the enabling relation $\vdash$. A notion of conflict between events is naturally given if one cell can be filled by different values. The following generalization of concrete data structures considers only events. This time consistency is defined a priori.

Definition 2.2 ([Win87, Def. 1.1.1]). An *event structure* or an *evs*, for short, is a triple $\mathcal{E} = (E, Cons, \vdash)$ such that

(1) E is a countable set of *events*,

(2) $Cons \subseteq Fin(E)$ is non-empty, called the collection of *consistent sets*, and whenever $A \subseteq B$ and $B \in Cons$, then $A \in Cons$,

(3) $\vdash \subseteq Cons \times E$, called the *enabling relation*, and whenever $A \vdash e$, $A \subseteq B$ and $B \in Cons$, then $B \vdash e$.

A subset X of E is called a *state* of $\mathcal{E}$, if the following two conditions are satisfied:

(1) *consistency*: for all finite $A \subseteq X$ we have $A \in Cons$,

(2) *safety*: for all $e \in X$ there is a sequence $e_0, \ldots, e_n \in X$ such that $e_n = e$ and for all $i \leq n$ we have $\{e_0, \ldots, e_{i-1}\} \vdash e_i$.

The set of all states of an evs $\mathcal{E}$, ordered by inclusion, is denoted by $(D(\mathcal{E}), \subseteq)$ and called the *event domain associated with* $\mathcal{E}$. If $(D, \leq)$ is isomorphic to $(D(\mathcal{E}), \subseteq)$, we say $(D, \leq)$ is generated by $\mathcal{E}$.

We note that it is easy to find for a given cds $\mathcal{M}$ an evs $\mathcal{E}$ such that $(D(\mathcal{M}), \subseteq) \cong (D(\mathcal{E}), \subseteq)$ (*cf.* [Cur93, p. 179],[Dro89, Prop. 2.8]).

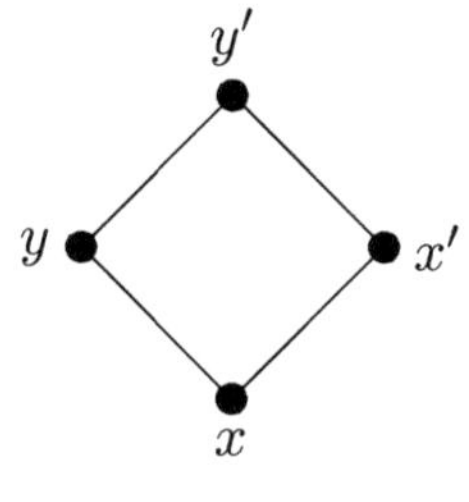

Figure 8.1. $[x,y] \prec [x',y']$

Next we wish to characterize the posets $(D, \leq)$ occurring as event domains $(D(\mathcal{E}), \subseteq)$ or concrete domains $(D(\mathcal{M}), \subseteq)$. We summarize some standard notation. Let $(D, \leq)$ be a domain, *i.e.* an algebraic complete partial order (cpo) with a least element. We denote the set of all *compact* (or isolated or finite) elements of $(D, \leq)$ by D^0. We say that $(D, \leq)$ is a *Scott-domain* if each non-empty subset of D which is bounded above in D has a supremum in D. If additionally D^0 is countable, we speak of an *ω-Scott-domain.* For any $x, y \in D^0$ with $x < y$ we write $x \prec y$ if there is no $z \in D$ with $x < z < y$. We denote such a pair by $[x, y]$ and call it a *prime interval* of D. The collection of all prime intervals of D is denoted by Int_D. For two prime intervals $[x, y], [x', y']$ of D we put $[x, y] \prec [x', y']$ if $x' \neq y$, $x \prec x'$ and $y \prec y'$, *cf.* Figure 8.1. We let $\succ\prec$ denote the smallest equivalence relation on Int_D containing $\prec$. If $[x, y] \succ\prec [x', y']$, we call the prime intervals *equivalent.* If $[x, y] \in Int_D$, let $[x, y]_{\succ\prec}$ denote the equivalence class of $[x, y]$ with respect to $\succ\prec$. Moreover, we write $x \uparrow y$ if $\{x, y\}$ has an upper bound, and $x \not\uparrow y$ otherwise. In case it exists, the supremum of x and y is denoted by $x \vee y$.

Definition 2.3 ([Dro89, Def. 2.2]). An ω-Scott-domain $(D, \leq)$ is called an *event domain,* if it satisfies the following conditions for any $x, x', y, y', z \in D^0$:

(F) $\{d \in D : d \leq x\}$ is finite;

(C) if $x \prec y$, $x \prec z$, $y \neq z$ and $y \uparrow z$, then $y \prec y \vee z$ and $z \prec y \vee z$;

(I) $[x, y] \succ\prec [x', y']$ and $x \leq x'$ imply $y \leq y'$.

Condition (F) says that compact elements in event domains are meant to stand for finite amounts of information. Condition (C) is a well-known condition from lattice theory called upper semi-modularity. Condition (I) contains as an instance:

(R) $[x, y] \succ\!\prec [x, z]$ implies $y = z$.

Now let $\mathcal{E}$ be an evs. Trivially, the empty set is the smallest state of $\mathcal{E}$. In $(D(\mathcal{E}), \subseteq)$ suprema are unions and finite states of $\mathcal{E}$ are exactly the compact elements. Hence property (F) holds. If $x \prec y$ in $(D(\mathcal{E}), \subseteq)$, there is $e \in E$ with $y = x \,\dot\cup\, \{e\}$. This implies property (C). Moreover, $[x, y] \succ\!\prec [x', y']$ and $y = x \,\dot\cup\, \{e\}$ imply $y' = x' \,\dot\cup\, \{e\}$. Hence, $(D(\mathcal{E}), \subseteq)$ satisfies condition (I).

Proposition 2.1 ([Dro89]). *Let $\mathcal{E}$ be an evs. Then $(D(\mathcal{E}), \subseteq)$ is an event domain whose compact elements are precisely the finite states of $\mathcal{E}$.*

Since any cds $\mathcal{M}$ can easily be regarded as an evs, $(D(\mathcal{M}), \subseteq)$ will also be an event domain. Moreover, $(D(\mathcal{M}), \subseteq)$ has additional properties. The following notion was introduced by Kahn and Plotkin.

Definition 2.4 (*cf.* [KP93, Def. 5.7]). An ω-Scott-domain $(D, \leq)$ is called a *concrete domain*, if it satisfies the conditions (F), (C), (R) and, moreover, for any $x, y, z \in D^0$:

(Q) if $x \prec z$, $x < y$ and $z \not\uparrow y$, then exists a uniquely determined $z' \in D^0$ such that $x \prec z' \leq y$ and $z \not\uparrow z'$.

It can be shown that a concrete domain also satisfies property (I), thus is an event domain. We include the argument why $(D(\mathcal{M}), \subseteq)$ satisfies condition (Q). Two states y and z of $\mathcal{M}$ do not have an upper bound iff there are $(c, v) \in y$ and $(c, v') \in z$ where $v \neq v'$. If now $x \prec z$, $x < y$ and $z \not\uparrow y$ in $(D(\mathcal{M}), \subseteq)$ it follows $z = x \,\dot\cup\, \{e\}$ for some $e = (c, v) \in E$. Furthermore, there is $e' = (c, v') \in y$ with $v' \neq v$. Hence, $z' = x \,\dot\cup\, \{e'\}$ is the unique element with $x \prec z' \leq y$ and $z' \not\uparrow z$.

Proposition 2.2 ([KP93, Thm. 6.22]). *Let $\mathcal{M}$ be a cds. The poset of states $(D(\mathcal{M}), \subseteq)$ is a concrete domain whose compact elements are precisely the finite states of $\mathcal{M}$.*

The connection between concrete and event domains is established by the following proposition.

Proposition 2.3 (*cf.* [Cur93, Prop. 2.2.11]). *A concrete domain is an event domain satisfying axiom*

(W) *for any $x, x', x'', y, y' \in D^0$ with $[x, x'] \succ\!\prec [y, y']$, $x \prec x''$ and $x' \not\uparrow x''$ there exists $y'' \in D^0$ such that $[x, x''] \succ\!\prec [y, y'']$ and $y' \not\uparrow y''$.*

Let $(D, \leq)$ be an event domain. A *chain* from x to y is a sequence $x_0, \ldots, x_n$ in D such that $x = x_0$, $y = x_n$ and $x_i \prec x_{i+1}$ for each

$i = 0, \ldots, n-1$. If $x, y \in D^0$ and $x \leq y$, then there is a chain from x to y because of property (F). For $z \in D$, we let

$$I(z) = \{[x,y] : y \leq z\},$$
$$s(z) = \{[x,y]_{\succ\!\prec} : y \leq z\}.$$

Proposition 2.4 ([Dro89]). *Let $(D, \leq)$ be an event domain.*
(a) *Let $x \in D^0$, and let $\bot = x_0 \prec x_1 \prec \ldots \prec x_n = x$ be a chain from $\bot$ to x. Then $s(x) = \{[x_{i-1}, x_i]_{\succ\!\prec} : i = 1, \ldots, n\}$.*
(b) *For any $x, y \in D$ we have $s(x) \subseteq s(y)$ iff $x \leq y$.*

Especially by the use of Proposition 2.4 and Proposition 2.3 one can show the following two representation theorems.

Theorem 2.5 ([KP93, Thm. 7.1]). *If $(D, \leq)$ is a concrete domain, then $(D, \leq) \cong (D(\mathcal{M}), \subseteq)$ for some cds $\mathcal{M}$.*

Theorem 2.6 ([Dro89, Thm. 2.7]). *Let $(D, \leq)$ be an event domain. Then $(D, \leq) \cong (D(\mathcal{E}), \subseteq)$ for some evs $\mathcal{E}$.*

Here in general neither the evs $\mathcal{E}$ nor the cds $\mathcal{M}$ are uniquely determined by $(D, \leq)$, and this is the starting point of this paper. We will show that the class of all evs generating a fixed event domain or all cds generating a fixed concrete domain, respectively, can be well structured.

3. Congruences on prime intervals and associated structures

Any domain $(D(\mathcal{E}), \subseteq)$ or $(D(\mathcal{M}), \subseteq)$ generated by an evs or a cds carries a natural labeling of its prime intervals, as indicated in the discussion before Proposition 2.1. This natural labeling induces a relation on Int_D with certain properties inherited from the definition of a state of an evs or a cds, respectively. In this section, we define such relations for any event and concrete domain, respectively, and then, in turn, associate generating structures by means of these relations.

In the sequel, we use the notation $x \prec \{x' \not\uparrow x''\}$ whenever $x, x', x'' \in D^0$ and $x \prec x'$, $x \prec x''$ with $x' \not\uparrow x''$.

Definition 3.1. **(a)** Let $(D, \leq)$ be an event domain. An equivalence relation $\equiv_e$ on Int_D is called an *event congruence* if

(1) $\succ\!\prec \; \subseteq \; \equiv_e$, and
(2) whenever $x, y \in D^0$ are such that for any $[w, w'] \in I(x)$ there is $[z, z'] \in I(y)$ with $[w, w'] \equiv_e [z, z']$, then $x \leq y$.

(b) Let $(D, \leq)$ be a concrete domain. An equivalence relation $\equiv_c$ on Int_D is a *concrete congruence* if conditions (1) and (2) hold and, moreover,

(3) whenever $x \prec \{x' \not\uparrow x''\}$, $y \prec y'$ and $[x, x'] \equiv_c [y, y']$, then there is some $y'' \in D^0$ with $y \prec \{y' \not\uparrow y''\}$ and $[x, x''] \equiv_c [y, y'']$.

Using Propositions 2.4 (b) and 2.3 it is easy to show that $\succ\!\prec$ is an event congruence for event domains and a concrete congruence for concrete domains. Moreover, let $\mathcal{E}$ be an evs and $(D(\mathcal{E}), \subseteq)$ the generated event domain. We define the *induced event congruence* $\equiv_{\mathcal{E}}$ by letting $[x, x'] \equiv_{\mathcal{E}} [y, y']$ if $x' = x \cup \{e\}$ and $y' = y \cup \{e\}$ for some event $e \in E$. Then $\equiv_{\mathcal{E}}$ is an event congruence on $(D(\mathcal{E}), \subseteq)$ as is easy to check. In the same way, one defines the *induced concrete congruence* $\equiv_{\mathcal{M}}$ for generated concrete domains $(D(\mathcal{M}), \subseteq)$ which yields a concrete congruence on this domain.

Our aim is the construction of an evs or cds from a given domain using particularly defined congruences. In order to construct the cells of a cds we introduce one more relation.

Definition 3.2. Let $(D, \leq)$ be a concrete domain and $\equiv_c$ a concrete congruence on it. Let B denote the equivalence classes of $\equiv_c$. Define a binary relation $\# \subseteq B \times B$ as follows: for $b_1, b_2 \in B$ let $b_1 \# b_2$ if either $b_1 = b_2$ or there are x, x_1, x_2 with $x \prec \{x_1 \not\uparrow x_2\}$ and $[x, x_1] \in b_1$, $[x, x_2] \in b_2$. Then $\#$ is called the *conflict relation* on $(D, \leq)$ with respect to $\equiv_c$.

Let $\mathcal{M}$ be a cds, $(D(\mathcal{M}), \subseteq)$ its generated concrete domain endowed with the induced concrete congruence $\equiv_{\mathcal{M}}$, and b_1, b_2 two different equivalence classes of $\equiv_{\mathcal{M}}$. Then $b_1 \# b_2$ iff there are representatives $[x, x_1] \in b_1$ and $[x.x_2] \in b_2$ such that $x_1 = x \cup \{(c, v)\}$, $x_2 = x \cup \{(c, v')\}$, and $v \neq v'$. Obviously, this is an equivalence relation. In general we also obtain:

Lemma 3.1. *Every conflict relation $\#$ on a concrete domain $(D, \leq)$ with respect to a concrete congruence $\equiv_c$ is an equivalence relation.*

Proof. Obviously, $\#$ is reflexive and symmetric. We show transitivity. Let $b_1, b_2, b_3 \in B$ such that $b_1 \# b_2$ and $b_2 \# b_3$. If $b_1 = b_2$ or $b_2 = b_3$, we are done. Otherwise, there are x, x_1, x_2 such that $x \prec \{x_1 \not\uparrow x_2\}$ and $[x, x_1] \in b_1$, $[x, x_2] \in b_2$. Similarly we have $y \prec \{y_2 \not\uparrow y_3\}$ such that $[y, y_2] \in b_2$, $[y, y_3] \in b_3$. By condition (3) of Definition 3.1 there is $x_3 \in D^0$ such that $x \prec \{x_2 \not\uparrow x_3\}$ and $[x, x_3] \in b_3$, in particular $x_2 \neq x_3$. If $x_1 = x_3$ we have $b_1 = b_3$ and hence $b_1 \# b_3$. Let $x_1 \neq x_3$. If $x_1 \uparrow x_3$, then $x_1 \prec (x_1 \vee x_3)$ and $x_3 \prec (x_1 \vee x_3)$ by property (C). Since $x_1 \not\uparrow x_2$, also $x_2 \not\uparrow (x_1 \vee x_3)$. By property (Q) there is a unique $x^\star \in D^0$ such that $x \prec x^\star \leq (x_1 \vee x_3)$ and $x^\star \not\uparrow x_2$. But both x_1 and x_3 fulfill this condition. This contradicts property (Q). Hence, $x_1 \not\uparrow x_3$ and therefore $b_1 \# b_3$. This shows transitivity of $\#$. □

By $[I(x)]_{\equiv_c}$ we denote as usual the set $\{[y,z]_{\equiv_c} : [y,z] \in I(x)\}$.

Proposition 3.2. *Let $(D, \leq)$ be a concrete domain and $\equiv_c$ a concrete congruence on it. Let $x \in D$ and $b_1, b_2 \in [I(x)]_{\equiv_c}$. Then $b_1 \# b_2$ implies $b_1 = b_2$.*

Proof. Assume $b_1 \neq b_2$. By Proposition 2.4(a) we may assume that there are $x_1 \prec x_2 \leq x_3 \prec x_4 \leq x$ such that $[x_1, x_2] \in b_1$ and $[x_3, x_4] \in b_2$. Since $b_1 \# b_2$, there is some $x_2' \in D^0$ with $[x_1, x_2'] \in b_2$ and $x_2 \not\uparrow x_2'$. Now we have $[I(x_2')]_{\equiv_c} \subseteq [I(x_4)]_{\equiv_c}$. Therefore $x_2' \leq x_4$. Since also $x_2 \leq x_4$, this contradicts $x_2 \not\uparrow x_2'$. Hence $b_1 = b_2$. □

Now we show how to construct from a given event domain with an event congruence an evs that generates the domain and yields the same congruence. We also provide a corresponding construction of cds from concrete domains with a concrete congruence. In the particular case that the given event or concrete congruence, respectively, equals $\succ\!\prec$, this is precisely the construction of [Dro89] and very similar to the construction of [KP93].

Construction 3.3. Let $(D, \leq)$ be an event domain with a given event congruence $\equiv_e$. Let E be the set of equivalence classes of $\equiv_e$. We define an *associated evs* $\boldsymbol{es}(D, \leq, \equiv_e) = (E, \mathit{Cons}, \vdash)$ as follows:

(1) for $A \subseteq E$, we let $A \in \mathit{Cons}$ if $A \subseteq [I(x)]_{\equiv_e}$ for some $x \in D^0$,

(2) for any $A \in \mathit{Cons}$ and $e \in E$, we let $A \vdash e$ if there are $x, y \in D^0$ with $[I(x)]_{\equiv_e} \subseteq A$ and $e = [x,y]_{\equiv_e}$.

Construction 3.4. Let $(D, \leq)$ be a concrete domain with a given concrete congruence $\equiv_c$. Let B be the set of equivalence classes of $\equiv_c$. Moreover, let $[b]_\#$ be the equivalence class of $b \in B$ with respect to $\#$. The *associated cds* $\boldsymbol{cds}(D, \leq, \equiv_c) = (C, V, E, \vdash)$ is defined as follows:

(1) $C = \{[b]_\# : b \in B\}$,

(2) $V = B$,

(3) $E = \{([b]_\#, b) : b \in B\}$,

(4) for any $A \subseteq B$ and $b \in B$, we let $\{([a]_\#, a) : a \in A\} \vdash [b]_\#$ if there are $x, y, z \in D^0$ such that $[I(x)]_{\equiv_c} \subseteq A \subseteq [I(z)]_{\equiv_c}$ and $b = [x,y]_{\equiv_c}$.

We will call two event (resp. concrete) congruences $\equiv_1$ and $\equiv_2$ on two given event (resp. two concrete) domains *isomorphic* if there is an order-isomorphism of the domains which preserves the considered congruences in both directions, *i.e.*, yields a bijective mapping between the sets of equivalence classes. We write $\equiv_1 \cong \equiv_2$.

Theorem 3.5. **(a)** *The associated evs* $\mathcal{E} = \boldsymbol{es}(D, \leq, \equiv_e)$ *is an evs generating* $(D, \leq)$. *Moreover, the given event congruence* $\equiv_e$ *and the induced event congruence* $\equiv_{\mathcal{E}}$ *are isomorphic.*
(b) *The associated cds* $\mathcal{M} = \boldsymbol{cds}(D, \leq, \equiv_c)$ *is a cds generating* $(D, \leq)$. *The given concrete congruence* $\equiv_c$ *and the concrete congruence* $\equiv_{\mathcal{M}}$ *induced by* $\mathcal{M}$ *are isomorphic.*

Proof. We prove only part (b) in detail and indicate how to derive the easier part (a).

Let $\mathcal{M} = \boldsymbol{cds}(D, \leq, \equiv_c)$. Trivially, every cell $[b]_\#$ is filled and because of property (F) we have $\vdash \subseteq Fin(E) \times C$. By assumption $(D, \leq)$ is an ω-Scott-domain. Hence, C and V are countable. Thus, $\mathcal{M}$ is a cds. Let $(D(\mathcal{M}), \subseteq)$ be the concrete domain generated by $\mathcal{M}$. We define a mapping $\psi : (D, \leq) \to (D(\mathcal{M}), \subseteq)$ by $\psi(x) := \{([b]_\#, b) : b \in [I(x)]_{\equiv c}\}$, and we will show that ψ is an isomorphism. Indeed, ψ is well defined: Let $x \in D$. If $([b_1]_\#, b_1), ([b_2]_\#, b_2) \in \psi(x)$ with $[b_1]_\# = [b_2]_\#$, we have $b_1 \# b_2$. Furthermore, $b_1, b_2 \in [I(x)]_{\equiv c}$ by definition of ψ. By Proposition 3.2 we get $b_1 = b_2$. Hence, $\psi(x)$ is consistent. If $([b]_\#, b) \in \psi(x)$, there are $y \prec y' \leq x$ such that $[y, y']_{\equiv c} = b$. Choose a chain $\bot = y_0 \prec y_1 \prec \ldots \prec y_n = y$ from $\bot$ to y in $(D, \leq)$. Put $y_{n+1} = y'$ and $b_i = [y_{i-1}, y_i]_{\equiv c}$ for $i = 1, \ldots, n+1$. Then $\{b_j : j \leq i\} = [I(y_i)]_{\equiv c}$ and $\{([b_j]_\#, b_j) : j < i\} \vdash [b_i]_\#$ for $i = 1, \ldots, n+1$. Hence, $\psi(x)$ is safe and so $\psi(x) \in D(\mathcal{M})$.

Next we show that $\psi|_{D^0} : (D^0, \leq) \to (D^0(\mathcal{M}), \subseteq)$ is an isomorphism. Let $x, y \in D^0$. If $x \leq y$, clearly $I(x) \subseteq I(y)$, so $\psi(x) \subseteq \psi(y)$. Conversely, $\psi(x) \subseteq \psi(y)$ implies $[I(x)]_{\equiv c} \subseteq [I(y)]_{\equiv c}$ and by the properties of $\equiv_c$ we have $x \leq y$. This shows injectivity as well. It remains to show that $\psi|_{D^0}$ is onto. We proceed by induction on the number of elements of finite states of $\mathcal{M}$. Trivially, $\psi(\bot_D) = \bot_{D(\mathcal{M})}$. If $A \in D^0(\mathcal{M})$, choose a chain $\bot = A_0 \prec A_1 \prec \ldots \prec A_n = A$ from $\bot$ to A in $(D(\mathcal{M}), \subseteq)$. Then $A = A_{n-1} \,\dot\cup\, \{([b]_\#, b)\}$ for some $b \in B$, and $A_{n-1} = \psi(y)$ for some $y \in D^0$ by induction hypothesis. Also, $[b]_\#$ is enabled in A_{n-1}, so there are $x, x' \in D^0$ such that $\psi(x) \subseteq A_{n-1} = \psi(y)$ and $[x, x']_{\equiv c} = b$. Hence $x \leq y$ as shown above. Assume $x' \not\uparrow y$. By property (Q) there is a unique $x'' \in D^0$ such that $x \prec x'' \leq y$ and $x' \not\uparrow x''$. For $a = [x, x'']_{\equiv c}$ we have $a \neq b$ and $a \# b$. But $a \in [I(y)]_{\equiv c}$, so $([a]_\#, a) \in \psi(y)$. Then the consistency of A implies $a = b$ which contradicts $a \neq b$. Hence, $x' \uparrow y$. Now choose a chain $x = x_0 \prec x_1 \prec \ldots \prec x_n = y$ from x to y in $(D, \leq)$. Since $b \notin [I(y)]_{\equiv c}$ we have $x' \not\leq y$, so $x' \neq x_1$. But $x' \uparrow y$ implies $x' \uparrow x_1$. Applying property (C), there exists $x' \vee x_1$ such that $x' \prec x' \vee x_1$, $x_1 \prec x' \vee x_1$ and $[x_1, x' \vee x_1]_{\equiv c} = b$, *cf.* Figure 8.2. Applying this argument along the chain $x = x_0 \prec x_1 \prec \ldots \prec x_n = y$

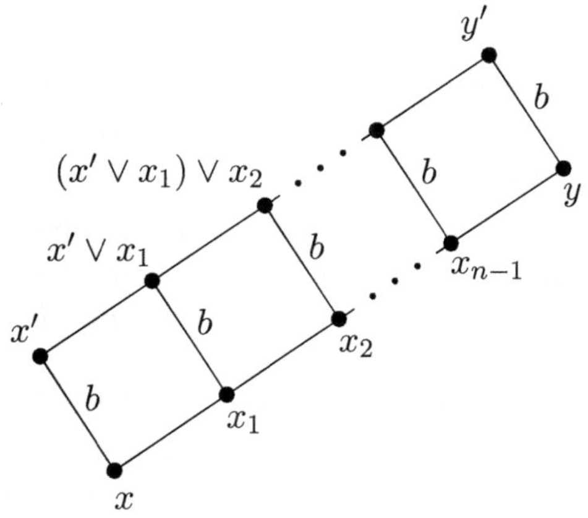

Figure 8.2. Showing surjectivity of ψ

we finally get some $y' \in D^0$ such that $y \prec y'$, $x' \leq y'$ and $[y, y']_{\equiv_c} = b$. This shows that $A = \psi(y) \cup \{([b]_\#, b)\} = \psi(y') \in \psi(D^0)$ as claimed.

Clearly, if $x \in D$, we have $\psi(x) = \bigcup\{\psi(x') : x' \in D^0, x' \leq x\}$. Hence $\psi : (D, \leq) \to (D(\mathcal{M}), \subseteq)$ is continuous and therefore an isomorphism. Moreover, it is easy to see that $[x, x'] \equiv_c [y, y']$ iff $[\psi(x), \psi(x')] \equiv_{\mathcal{M}} [\psi(y), \psi(y')]$ for all $x, x', y, y' \in D^0$. This proves part (b). The proof of part (a) follows mainly the same lines. But instead of property (Q) we can use the given notion of consistency to show that x' and y (*cf.* Figure 8.2) have an upper bound. □

The associated structures of Constructions 3.3 and 3.4 have additional properties: they do not contain 'superfluous' elements and their consistency and enabling relations are restricted as well. To make this precise we call an evs $\mathcal{E} = (E, Cons, \vdash)$ *trim*, if

(1) each event $e \in E$ is contained in some state X of $\mathcal{E}$,

(2) whenever $A \in Cons$, there is a state X of $\mathcal{E}$ with $A \subseteq X$, and

(3) whenever $A \vdash e$, there is a state Y of $\mathcal{E}$ with $Y \subseteq A$ such that $Y \cup \{e\}$ is a state of $\mathcal{E}$.

A cds $\mathcal{M} = (C, V, E, \vdash)$ is *trim*, if

(1) each value $v \in V$ is contained in some event $e \in E$,

(2) each event $e \in E$ is an element of some state X of $\mathcal{M}$, and

(3) for every finite set $A \subseteq E$ and $c \in C$ we have $A \vdash c$ iff there are finite states X, Z such that $X \subseteq A \subseteq Z$ and for all $v \in V$ with $(c, v) \in E$ the set $Y = X \dot\cup \{(c, v)\}$ is a state of $\mathcal{M}$.

Here, we use the word *trim* similarly to the notion of trim automata in automata theory.

To state our next result correctly we have to introduce morphisms between evs and between cds. Let π_1 denote the projection onto the first component of a tuple.

Definition 3.3. **(a)** Let $\mathcal{E}$ and $\mathcal{E}'$ be two evs. A function $g : E \to E'$ is called a *morphism* from $\mathcal{E}$ to $\mathcal{E}'$, if g preserves consistency and enabling. A bijective morphism $g : \mathcal{E} \to \mathcal{E}'$ is called an *isomorphism*, if consistency and enabling are also preserved by g^{-1}.
(b) Let $\mathcal{M} = (C, V, E, \vdash)$ and $\mathcal{M}' = (C', V', E', \vdash')$ be two cds. A pair of functions (g_C, g_E) with $g_C : C \to C'$ and $g_E : E \to E'$ is called a *morphism* from $\mathcal{M}$ to $\mathcal{M}'$, if the following conditions hold:

(1) $\pi_1 \circ g_E(c, v) = g_C \circ \pi_1(c, v) = g_C(c)$ for all $(c, v) \in E$;
(2) if $A \in Fin(E)$ and $c \in C$, then $A \vdash c$ implies $g_E(A) \vdash' g_C(c)$;
(3) if X is a state of $\mathcal{M}$ and $(c_1, v_1), (c_2, v_2) \in X$ such that $g_C(c_1) = g_C(c_2)$, then $g_E(c_1, v_1) = g_E(c_2, v_2)$.

If g_C and g_E are bijective, and if in condition (2) also the inverse implication holds, then (g_C, g_E) is an *isomorphism*.

So, morphisms preserve consistency and enabling, and for cds, moreover, also the cell of the events. A morphism between cds is uniquely determined by the event mapping g_E, thus g_C is introduced only for technical reasons. We note that states of evs and of cds are preserved under the respective morphisms. Therefore, each morphism $g : \mathcal{E} \to \mathcal{E}'$ induces a continuous mapping $\overline{g} : (D(\mathcal{E}), \subseteq) \to (D(\mathcal{E}'), \subseteq)$ by $\overline{g}(X) = g(X)$ for any state X of $\mathcal{E}$. Then $\overline{g}$ preserves the least and the compact elements. The same holds true for morphisms between cds.

The proof of the following theorem is rather technical and, therefore, omitted. The result was shown in detail in [BDM01].

Theorem 3.6. *The associated evs $\mathcal{E} = \boldsymbol{es}(D, \leq, \equiv_e)$ of Construction 3.3 and the associated cds $\mathcal{M} = \boldsymbol{cds}(D, \leq, \equiv_c)$ of Construction 3.4 are trim. Moreover, $\mathcal{E}$ is the unique trim evs, up to isomorphism, generating $(D, \leq)$ with $\equiv_{\mathcal{E}}$ isomorphic to $\equiv_e$, and $\mathcal{M}$ is, up to isomorphism, the unique trim cds generating $(D, \leq)$ with $\equiv_{\mathcal{M}}$ isomorphic to $\equiv_c$.*

Let $\mathcal{E}, \mathcal{E}'$ be two evs with underlying event sets E, E', respectively. We call $\mathcal{E}'$ a *trim kernel* of $\mathcal{E}$ if $\mathcal{E}'$ is trim, $E' \subseteq E$ and $D(\mathcal{E}) = D(\mathcal{E}')$, *i.e.*, $\mathcal{E}$ and $\mathcal{E}'$ have the same states. The analogous definition applies to a *trim kernel* of a cds $\mathcal{M}$. The following lemma shows that it is sufficient to consider only trim evs and cds.

Lemma 3.7. *For any evs $\mathcal{E}$ and any cds $\mathcal{M}$ there is a uniquely determined trim kernel $\mathcal{E}'$ or $\mathcal{M}'$, respectively.*

4. Maximal and minimal structures

In classical formal language theory each automaton can be reduced to a minimal one recognizing the same language. Similarly, we show in this section that the collection of all trim evs (resp. trim cds) generating a given event (resp. concrete) domain $(D, \leq)$ contains maximal and minimal structures. In particular, we maximize or minimize the number of events and for cds also the number of cells but not the number of values. At first we introduce a necessary tool.

Definition 4.1. We will call a morphism $g : \mathcal{E} \to \mathcal{E}'$, respectively $(g_C, g_E) : \mathcal{M} \to \mathcal{M}'$, a *reduction* if $g : E \to E'$, respectively $g_E : E \to E'$, is onto and the induced mapping $\overline{g}$ is an isomorphism of the considered domains.

In view of Theorem 3.6 and Lemma 3.7, we will restrict ourselves in the sequel to trim evs and cds. We will say in the following that *a property P holds* for two congruences $\equiv_1$ and $\equiv_2$ *up to isomorphism* if there are isomorphic congruences $\equiv_i' \cong \equiv_i$ for $i = 1, 2$ such that P holds for $\equiv_1'$ and $\equiv_2'$.

Lemma 4.1. *Let $\mathcal{E}$ and $\mathcal{E}'$ be two trim evs generating the same event domain $(D, \leq)$. Then there is a reduction $g : \mathcal{E} \to \mathcal{E}'$ iff $\equiv_\mathcal{E} \subseteq \equiv_{\mathcal{E}'}$ up to isomorphism. Similarly, for two trim cds $\mathcal{M}, \mathcal{M}'$ generating the same concrete domain $(D, \leq)$, there is a reduction $(g_C, g_E) : \mathcal{M} \to \mathcal{M}'$ iff $\equiv_\mathcal{M} \subseteq \equiv_{\mathcal{M}'}$ up to isomorphism.*

Proof. Let $\mathcal{E}, \mathcal{E}'$ be two trim evs generating $(D, \leq)$. Assume there is a reduction $g : \mathcal{E} \to \mathcal{E}'$. Hence, $\overline{g} : (D(\mathcal{E}), \subseteq) \to (D(\mathcal{E}'), \subseteq)$ is an isomorphism. Let X_i, Y_i be finite states of $(D(\mathcal{E}), \subseteq)$ with $X_i \prec Y_i$ for $i = 1, 2$. Then $[X_1, Y_1] \equiv_\mathcal{E} [X_2, Y_2]$ iff $Y_1 \setminus X_1 = Y_2 \setminus X_2 = \{e\}$ for some $e \in E$, and similar for $\equiv_{\mathcal{E}'}$. Therefore, we obtain $\equiv_\mathcal{E} \subseteq \equiv_{\mathcal{E}'}$ up to isomorphism.

Conversely, we assume $\equiv_\mathcal{E} \subseteq \equiv_{\mathcal{E}'}$ up to isomorphism. There are congruences $\equiv \cong \equiv_\mathcal{E}$ and $\equiv' \cong \equiv_{\mathcal{E}'}$ on $(D, \leq)$ such that $\equiv \subseteq \equiv'$. Let $\mathcal{E}_D = \boldsymbol{es}(D, \leq, \equiv)$ and $\mathcal{E}_D' = \boldsymbol{es}(D, \leq, \equiv')$. By Theorem 3.6, $\mathcal{E} \cong \mathcal{E}_D$ and $\mathcal{E}' \cong \mathcal{E}_D'$. Therefore, it suffices to show that there is a reduction from $\mathcal{E}_D$ to $\mathcal{E}_D'$. Now we define $g : E_D \to E_D'$ by $g : [x, y]_\equiv \mapsto [x, y]_{\equiv'}$ for any prime interval $[x, y]$ of $(D, \leq)$. Since $\equiv \subseteq \equiv'$, g is well defined. Moreover, g is onto. We show that g is a morphism from $\mathcal{E}_D$ to $\mathcal{E}_D'$. Let $A \subseteq E_D$ with $A \in \mathit{Cons}$. Due to Construction 3.3 there is some $x \in D^0$ with $A \subseteq [(I(x)]_\equiv$. Hence, $g(A) \subseteq [I(x)]_{\equiv'}$ and $g(A) \in \mathit{Cons}'$. Next, let $A \vdash e$ for $A \in \mathit{Cons}$ and $e \in E_D$. Again due to Construction 3.3 there are $x, y \in D^0$ with $[I(x)]_\equiv \subseteq A$ and $[x, y]_\equiv = e$. This implies immediately $[I(x)]_{\equiv'} \subseteq g(A)$ and $[x, y]_{\equiv'} = g(e)$. Hence, $g(A) \vdash' g(e)$.

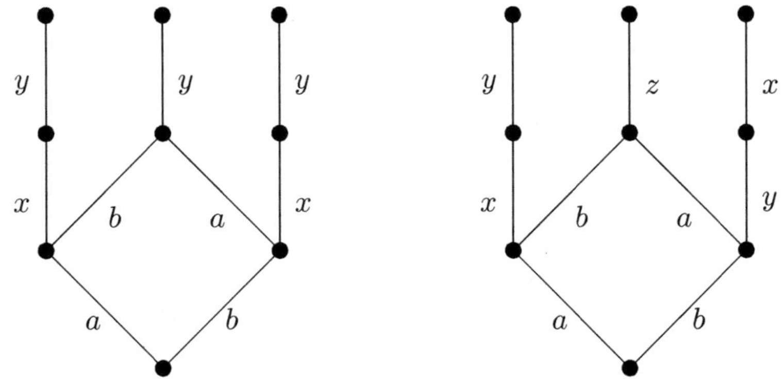

Figure 8.3. Two maximal event congruences of an event domain

Thus $g : \mathcal{E}_D \to \mathcal{E}'_D$ is a morphism and a reduction. The proof for cds follows the same lines. □

For any event domain $(D, \leq)$ we call $\mathcal{E}_D = \boldsymbol{es}(D, \leq, \asymp)$ the *canonical evs associated with* $(D, \leq)$. If $(D, \leq)$ is a concrete domain, $\mathcal{M}_D = \boldsymbol{cds}(D, \leq, \asymp)$ is the *canonical cds associated with* $(D, \leq)$. Due to Theorem 3.6 canonical structures are trim. Moreover, we show now:

Theorem 4.2. *Let $(D, \leq)$ be any event domain. The canonical evs $\mathcal{E}_D$ associated with $(D, \leq)$ is with regard to reduction the greatest trim evs among all trim evs generating $(D, \leq)$. The same holds true for concrete domains and trim cds.*

Proof. By Theorem 3.5, $\mathcal{E}_D$ generates $(D, \leq)$. Let $\mathcal{E}$ be any trim evs generating $(D, \leq)$ and $\equiv_{\mathcal{E}}$ be the event congruence on $(D, \leq)$ induced by $\mathcal{E}$. Of course, $\asymp \subseteq \equiv_{\mathcal{E}}$. Moreover, $\asymp$ is up to isomorphism the event congruence on $(D, \leq)$ induced by $\mathcal{E}_D$. By Lemma 4.1 there is a reduction from $\mathcal{E}_D$ to $\mathcal{E}$. To prove uniqueness, let $\mathcal{E}'$ be another trim evs generating $(D, \leq)$ and let $g : \mathcal{E}' \to \mathcal{E}_D$ be a reduction. Then $\equiv_{\mathcal{E}'} \subseteq \asymp$ up to isomorphism by Lemma 4.1. Hence, there are congruences $\equiv' \cong \equiv_{\mathcal{E}'}$ and $\equiv \cong \asymp$ on $(D, \leq)$ with $\equiv' \subseteq \equiv$. Since $\asymp \subseteq \equiv$ trivially, and order-isomorphisms preserve $\asymp$, we obtain $\equiv = \asymp$. Hence $\equiv' \subseteq \asymp$, and thus $\equiv' = \asymp$. By Theorem 3.6, $\mathcal{E}' \cong \boldsymbol{es}(D, \leq, \equiv') = \mathcal{E}_D$. □

We call a trim evs $\mathcal{E}$ *minimal*, if whenever $\mathcal{E}'$ is another trim evs and $g : \mathcal{E} \to \mathcal{E}'$ is a reduction, then g is an isomorphism. Minimal trim cds are defined analogously.

Theorem 4.3. *Let $\mathcal{E}$ be any trim evs generating some event domain $(D, \leq)$. Then $\mathcal{E}$ can be reduced to a minimal trim evs $\mathcal{E}_{min}$. Similarly, any trim cds $\mathcal{M}$ can be reduced to a minimal trim cds $\mathcal{M}_{min}$ generating the same concrete domain.*

Proof. Let $\equiv_{\mathcal{E}}$ be the event congruence on $(D, \leq)$ induced by $\mathcal{E}$. As it is easy to check, the union of any chain (with respect to inclusion) of event congruences of $(D, \leq)$ is again an event congruence. Hence, by Zorn's lemma there is a maximal event congruence $\equiv_{max}$ on$(D, \leq)$ containing $\equiv_{\mathcal{E}}$. We put $\mathcal{E}_{min} = \boldsymbol{es}(D, \leq, \equiv_{max})$. By Lemma 4.1 there is a reduction from $\mathcal{E}$ to $\mathcal{E}_{min}$. Now let $\mathcal{E}'$ be any trim evs and $g : \mathcal{E}_{min} \to \mathcal{E}'$ a reduction. Using Lemma 4.1, there is an event congruence $\equiv'$ on $(D, \leq)$ isomorphic to $\equiv_{\mathcal{E}'}$ with $\equiv_{max} \subseteq \equiv'$, and by the maximality of $\equiv_{max}$ we get $\equiv' = \equiv_{max}$. Therefore, $\equiv_{max} = \equiv_{\mathcal{E}'}$ up to isomorphism and thus $\mathcal{E}_{min} \cong \mathcal{E}'$ by Theorem 3.6. A precise analysis shows that the given reduction g is a composition of the isomorphisms indicated here and, therefore, itself an isomorphism as needed. The proof for cds follows exactly the same lines. □

Therefore, searching for minimal trim evs and cds means searching for maximal event and concrete congruences on the prime intervals of the considered domain. Then we can associate the evs or cds by using the Constructions 3.3 and 3.4. This is illustrated by the following two examples.

Example 4.4. Let $(D, \leq)$ be the event domain given in Figure 8.3. Now consider the indicated two event congruences $\equiv_i$ for $i = 1, 2$ where two prime intervals are equivalent iff they are labeled with the same letter. Thus the sets of equivalence classes are $E_1 = \{a, b, x, y\}$ and $E_2 = E_1 \cup \{z\}$, respectively. It is easy to check that both event congruences are maximal. Hence by Theorem 4.3, $\mathcal{E}_i = \boldsymbol{es}(D, \leq, \equiv_i)$ $(i = 1, 2)$ are two minimal trim evs generating $(D, \leq)$. Note that $|E_1| \neq |E_2|$.

Example 4.5. Let $(D, \leq)$ be the concrete domain given in Figure 8.4. Consider the two concrete congruences $\equiv_c$ and $\equiv_c'$ again indicated by the labelings in the figure: the set of equivalence classes is $B = B' = \{a, b, c, d, e\}$. It is easy to see that $\equiv_c$ and $\equiv_c'$ are maximal. Hence by Theorem 4.3, $\mathcal{M} = \boldsymbol{cds}(D, \leq, \equiv_c)$ and $\mathcal{M}' = \boldsymbol{cds}(D, \leq, \equiv_c')$ are two

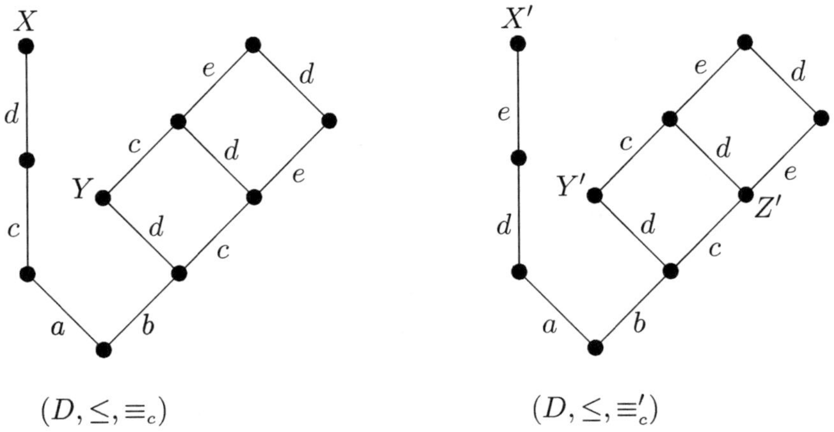

Figure 8.4. Two maximal concrete congruences of a concrete domain

minimal trim cds generating $(D, \leq)$. If we consider $[a]_\# = [b]_\#$ we get

$$C = C' = \{[a]_\#, [c]_\#, [d]_\#, [e]_\#\},$$
$$V = V' = \{a, b, c, d, e\},$$
$$E = E' = \{([a]_\#, a), ([b]_\#, b), ([c]_\#, c), ([d]_\#, d), ([e]_\#, e)\}.$$

Now we will show that $\mathcal{M}$ and $\mathcal{M}'$ are not isomorphic. Indeed, assume there is an isomorphism $(g_C, g_E) : \mathcal{M} \to \mathcal{M}'$. Let X, X', Y, Y', Z' be as shown in the figure. Interpreting X, Y and X', Y', Z' as states of $(D(\mathcal{M}), \subseteq)$ and $(D(\mathcal{M}'), \subseteq)$, respectively, we get $X' = \overline{g}(X)$ and therefore $g([d]_\#, d) = ([e]_\#, e)$. But $\overline{g}$ has to be an isomorphism of the concrete domains. This implies $Y' = \overline{g}(Y)$ or $Z' = \overline{g}(Y)$. Hence, $g([d]_\#, d) = ([d]_\#, d)$ or $g([d]_\#, d) = ([c]_\#, c)$ which contradicts $g([d]_\#, d) = ([e]_\#, e)$ as seen above. Hence, $\mathcal{M}$ and $\mathcal{M}'$ are not isomorphic.

In both examples, finding the maximal congruences was much easier than guessing the minimal generating evs or cds directly. In Example 4.4 we get two minimal evs with different numbers of events. In Example 4.5 the minimal cds have the same number of cells and events. They are not isomorphic because of the enabling relation. It remains an open question if minimal trim cds always have the same number of cells and events.

5. Conclusion

In this paper we explored the connection between evs and event domains and between cds and concrete domains, respectively. We obtained results about the existence of minimal structures generating a given event or concrete domain, respectively. Besides the open question concerning the number of events and cells of minimal cds generating the same domain, it would be interesting to have algorithms computing minimal evs and cds for a given event or concrete domain. Moreover, we can put the same questions for subclasses of the domains considered in this paper, e.g. distributive concrete domains and stable cds.

Last but not least, it would be nice to have the notion of reduction defined only by the evs or cds, respectively. Given two evs $\mathcal{E}$ and $\mathcal{E}'$, we would like to decide if $\mathcal{E}$ can be reduced to $\mathcal{E}'$ without considering the generated domains.

References

[BC82] G. Berry and P.L. Curien, Sequential algorithms on concrete data structures, *Theoret. Comp. Science* **20** (1982), 265–321.

[BD94] F. Bracho and M. Droste, Labelled domains and automata with concurrency, *Theoret. Comp. Science* **135** (1994), 289–318.

[BD97] F. Bracho and M. Droste, *From domains to event structures*, MATH–AL–Preprint, TU Dresden, 1997.

[BDM01] F. Bracho, M. Droste and I. Meinecke, *From domains to event structures and concrete data structures*, MATH-AL-Preprint, TU Dresden, 2001.
see `http://www.math.tu-dresden.de/~meinecke`

[Ber78] G. Berry, Stable models of typed λ-calculi, 5th Intern. Colloq. on *Automata, Languages and Programming*, Lecture Notes in Comp. Science **62**, Springer Berlin, New York, 1978, pp. 72–89.

[Cur93] P.L. Curien, *Categorical Combinators, Sequential Algorithms and Functional Programming*, second ed., Progress in Theoret. Comp. Science, Birkhäuser Boston, 1993.

[Dro89] M. Droste, Event structures and domains, *Theoret. Comp. Science* **68** (1989), 37–47.

[DS02] M. Droste and R.M. Shortt, From Petri nets to automata with concurrency, *Applied Categorical Structures* **10** (2002), 173–191.

[KP78] G. Kahn and G.D. Plotkin, *Domaines concretes*, Rapport de Recherche no. 336, IRIA, Paris, 1978.

[KP93] G. Kahn and G.D. Plotkin, Concrete domains, *Theoret. Comp. Science* **121** (1993), 187–277.

[Kus94] D. Kuske, *Modelle nebenläufiger Prozesse – Monoide, Residuensysteme und Automaten*, Dissertation, Universität Essen, 1994.

[Mei00] I. Meinecke, *Von konkreten Bereichen zu konkreten Datenstrukturen*, Diplomarbeit, TU Dresden, 2000.

[NPW81] M. Nielsen, G. Plotkin, and G. Winskel, Petri nets, event structures and domains, part i, *Theoret. Comp. Science* **13** (1981), 85–108.

[Sco82] D. Scott, Domains for denotational semantics, 9th Intern. Colloq. on *Automata, Languages and Programming*, Lecture Notes in Comp. Science **140**, Springer Berlin, 1982, pp. 577–613.

[Win81] G. Winskel, *Events in computation*, Ph.D. thesis, Edinburgh, 1981.

[Win87] G. Winskel, Event structures, *Petri Nets: Applications and Relationships to Other Models of Concurrency* (W. Brauer, W. Reisig, and G. Rozenberg, eds.), Lecture Notes in Comp. Science **255**, Springer Berlin, New York, 1987, pp. 325–392.

[WN95] G. Winskel and M. Nielsen, Models for concurrency, In: *Handbook of Logic in Computer Science* (S. Abramsky, D.M. Gabbay, and T.S.E. Maibaum, eds.), vol. 4, Oxford University Press, 1995, pp. 1–148.

Chapter 9

A NOTE ON STRONGLY FINITE SEQUENT STRUCTURES*

Dieter Spreen
Theoretische Informatik, Fachbereich Mathematik
Universität Siegen
57068 Siegen, Germany
spreen@informatik.uni-siegen.de

Ralf Greb
Theoretische Informatik, Fachbereich Mathematik
Universität Siegen
57068 Siegen, Germany
ralfgreb@informatik.uni-siegen.de

Abstract Strongly finite sequent structures have been introduced by Zhang as a logic-oriented representation of SFP domains. In this paper we extend his definition by demanding the existence of a distinguished element with the properties of a tautology. This is no restriction compared with Zhang's approach, but has several advantages. Moreover, we show that the entailment relation of every strongly finite sequent structure can be derived from a preorder. A close connection between strongly finite sequent structures and certain preorders is exhibited. It helps clarifying the correspondence between strongly finite sequent structures and SFP domains. To every domain construction used in programming language semantics there is a similar construction for preorders from which the corresponding construction for sequent structures can be obtained in a uniform way.

Keywords: SFP domains, sequent structures, logic of domains

*This research has partially been supported by the DFG-RFFI grant "Domain-like Structures for Semantics and Computability"

GQ Zhang et al. (eds.), Domain Theory, Logic and Computation, 179-196.

1. Introduction

Domains introduced by Dana Scott [Sc70] have turned out to be an important mathematical tool for the formal description of programming language semantics. A domain is a partially ordered set with least element, in which every directed subset has a least upper bound. Moreover, every element can be approximated by special basic elements that satisfy a finiteness condition.

In order to provide a more computer science oriented approach to domain theory several representations of domains by logical calculi have been discussed in the literature. They are more concrete than partial orders. For domains in which also every bounded subset has a least upper bound, i.e. for Scott domains, Scott [Sc82] showed that they can be generated by information systems. An information system consists of a set of tokens, or propositions, together with an entailment relation between finite consistent sets of propositions and propositions, as well as a predicate specifying the consistent finite token sets.

Scott domains are used in many applications, but unfortunately the collection of such domains is not closed under all constructions needed in programming language semantics. This is, however, the case with the larger class of SFP domains introduced by Plotkin [Pl76].

In his thesis Zhang [Zh91] presented a logic oriented approach for the representation of SFP domains. The structures he considers consist of a set of propositions or tokens endowed with an entailment relation between finite sets of tokens, which satisfies the structural rules of Gentzen's sequent calculus and some additional requirements. They are called strongly finite sequent structures.

A mathematically very appealing systematic approach to connecting logic and domains has been developed by Abramsky [Ab91, AJ94] on the basis of Stone duality. As is well known this duality relates a topological space to its lattice of open sets. In the case of domains open sets can be represented as unions of compact-open sets. Since this representation is not unique, instead of lattices one has to consider preordered structures in which every pair of elements has an infimum and a supremum.

Compact-open sets are finite unions of principal filters generated by the finite (compact) elements of a domain. These filters are exactly the tokens of the sequent structures corresponding to SFP domains.

In this paper we consider sequent structures which contain a distinguished proposition having the properties of a tautology. Doing so has several advantages. If follows e.g. that any compact element in the domain associated with a sequent structure is generated by a token. If one works without such a distinguished token, the least element of the domain is the empty set, which leads to unnecessary case distinctions.

As we will see, the entailment relation of a sequent structure induces a preorder between the tokens. Moreover, any entailment relation can be generated from a preorder in a uniform way. We will thus work with preorders instead of entailment relations. Preorders can be considered as an intermediate level between sequence structures and domains by which their relationship becomes very transparent. The conditions a preorder has to satisfy so that the induced sequent structure is strongly finite correspond to those considered in [AJ94], but note that the preordered structure considered in that paper is more complex than the one studied here.

If, in the construction of the domain associated with a sequent structure, one uses how the entailment relation is generated from a preorder, then it turns out that the domain elements are exactly the filters of the preorder. Thus, this construction corresponds to the ideal completion of the dual preorder.

A further advantage of using preorders instead of entailment relations is that the usual domain constructions like products, function spaces and powerdomains can be reflected by preorder constructions in a perspicuous way. Note that in the function space case we will present a construction corresponding to the full continuous function space. In [Zh91] only the strict function space was considered.

The paper is organized as follows. Section 2 contains basic definitions and results from domain theory. In Section 3 the usual constructions of SFP domains are shortly reviewed. A slightly different construction of the Plotkin powerdomain is given, which solves a problem appearing in the approach displayed in [Zh91]. Strongly finite sequent structures and their relation to SFP domains are presented in Section 4. In Section 5 the above mentioned correspondence between strongly finite sequence structures and certain preorders with greatest element is derived. Constructions on such preorders that are analogous to the domain constructions considered in Section 3 are discussed in Section 6. The corresponding constructions for sequent structures can be derived from these in a uniform and simple way. Final remarks appear in Section 7.

2. Basic definitions and results

Let $(D, \sqsubseteq)$ be a partial order with smallest element $\perp$. For a subset S of D, let $\downarrow S = \{ x \in D \mid (\exists y \in S) x \sqsubseteq y \}$ and $\uparrow S = \{ x \in D \mid (\exists y \in S) y \sqsubseteq x \}$. S is called *directed*, if it is not empty and every pair of elements in S has an upper bound in S. If every directed subset S of D

has a least upper bound $\bigsqcup S$ in D, D is said to be a *directed-complete* partial order (cpo). If D is not a set but rather a class satisfying the analogous condition for directed subclasses, D is called *large* cpo.

An element x of a cpo D is *compact* if for any directed subset S of D the relation $x \sqsubseteq \bigsqcup S$ always implies the existence of an element $u \in S$ with $x \sqsubseteq u$. We denote the set of compact elements of D by D^0. In case that for every $y \in D$ the set $\downarrow\{y\} \cap D^0$ is directed with least upper bound y, the cpo D is said to be *algebraic*. If, in addition, D^0 is countable, it is ω*-algebraic* or, as we prefer to say, a *domain*. D^0 is then called its *(algebraic) basis*. Standard references for domain theory are [GS90, Gu92, AJ94, SLG94, AC98].

Let D and E be cpo's. A map $f\colon D \to E$ is *continuous* if it is monotone and for any directed subset S of D, $f(\bigsqcup S)$ is the least upper bound of $f(S)$. The space of all continuous functions from D into E is denoted by $[D \to E]$. Endowed with the *pointwise order*, i.e., $f \sqsubseteq g$ if $f(x) \sqsubseteq_E g(x)$, for all $x \in D$, it is a cpo again.

In order to be a domain, $[D \to E]$ has to contain special elements by which all other elements can be approximated.

Definition 2.1. Let D and E be cpo's. For $d \in D^0$ and $e \in E^0$ the *single-step function* $(d \searrow e)\colon D \to E$ is defined by

$$(d \searrow e)(x) = \begin{cases} e & \text{if } d \sqsubseteq x, \\ \bot & \text{otherwise.} \end{cases}$$

As is well known [AC98], $(d \searrow e) \in [D \to E]^0$. Moreover, if D and E are algebraic then we have that any $f \in [D \to E]$ is the least upper bound of $\{\, (d \searrow e) \mid (d \searrow e) \sqsubseteq f \,\}$.

Unfortunately, this set is not directed in general, which implies that in general $[D \to E]$ is not algebraic, even if D and E are. The set is directed, however, if the domains satisfy some conditions concerning the minimal upper bounds of finite sets of domains elements.

An element u of a cpo D is called a *minimal upper bound* of a subset X of D, if u is an upper bound of X that is not strictly greater than any other upper bound of X. Let $\mathrm{MUB}(X)$ denote the set of minimal upper bounds of X. As is well known, in a domain minimal upper bounds of finite sets of compact elements are compact again [Ju89]. A set S of upper bounds of X is *complete* for X if for every upper bound v of X there is some $u \in S$ below v. Define $U(X)$ to be the union of all sets $\mathrm{MUB}(Y)$, where Y ranges over the finite subsets of X, and let $U^*(X)$ be the smallest subset of D that contains X and is closed under U.

Definition 2.2. A cpo D is an *SFP domain* if it is ω-algebraic and for every finite subset X of D^0, MUB(X) is complete for X and $U^*(X)$ is finite.

SFP domains have been introduced by Plotkin [Pl76]. Their collection is closed under all important constructions needed in programming language semantics such a products, powerdomains and function spaces.

Let D_1 and D_2 be SFP domains and for $i = 1, 2$ let $p_i \colon D_1 \times D_2 \to D_i$ denote the projection onto the i-th component. A subset T of $D_1^0 \times D_2^0$ is *joinable* if for any finite subset T' of T one has that

$$(\forall a \in \mathrm{MUB}(p_1 T'))(\exists b \in \mathrm{MUB}(p_2 T'))(a, b) \in T.$$

If T is a finite joinable subset of $D_1^0 \times D_2^0$ the least upper bound of $\{ (a \searrow b) \mid (a, b) \in T \}$ exists in $[D_1 \to D_2]$. It is called *step function.* Maps of this kind are exactly the compact elements in $[D_1 \to D_2]$.

3. Domain constructions

As has been mentioned above, the collection of all SFP domains is not only closed under the function space construction, but also under constructions like lifting, sum, product as well as several powerdomain constructions. We skip the definition of the first three constructions and denote their result by $D_\perp$, $D + E$ and $D \times E$.

The following preorders are usually considered on the powerset $\mathcal{P}(D)$ of D:

$$\begin{aligned} A \sqsubseteq_S B &\Leftrightarrow (\forall b \in B)(\exists a \in A) a \sqsubseteq b, \\ A \sqsubseteq_H B &\Leftrightarrow (\forall a \in A)(\exists b \in B) a \sqsubseteq b, \quad \text{and} \\ A \sqsubseteq_P B &\Leftrightarrow A \sqsubseteq_S B \wedge A \sqsubseteq_H B. \end{aligned}$$

The corresponding powerdomains, called the Smyth, Hoare and Plotkin powerdomain, respectively, are the ideal completions of the collection $M(D)$ of all finite nonempty subsets of D^0 with respect to these preorders. The disadvantage of this construction is that the powerdomains thus obtained have their elements in $\mathcal{P}(\mathcal{P}(D))$, not in $\mathcal{P}(D)$. For many purposes a more concrete presentation of powerdomains of D is needed in which the elements are subsets of D. This is achieved in [Zh91] by picking up the largest element (with respect to set inclusion) in each equivalence class induced by the preorder under consideration, ordering them by the preorder, and adding least upper bounds for increasing chains of such elements.

For a subset X of D set

$$\begin{aligned}\mathrm{Cl}_S(X) &= \uparrow X,\\ \mathrm{Cl}_H(X) &= \downarrow X, \quad \text{and}\\ \mathrm{Cl}_P(X) &= \mathrm{Cl}_S \cap \mathrm{Cl}_H(\mathrm{Cl}_L(X)),\end{aligned}$$

where $\mathrm{Cl}_L(X)$ is the closure of X in the Lawson topology. This topology is generated by the subbase of all sets $\uparrow\{u\}$ and $D \setminus \uparrow\{v\}$, for $u, v \in D^0$. Note that $\mathrm{Cl}_L(X) = X$, if X is finite.

Theorem 3.1. *Let D be a domain. Then the structure $(\mathcal{P}_S(D), \sqsubseteq_S)$, where $\mathcal{P}_S(D)$ consists of all sets $\mathrm{Cl}_S(X)$ with $X \in M(D)$ together with all intersections $\bigcap_{i\in\omega} \mathrm{Cl}_S(X_i)$ of chains $\mathrm{Cl}_S(X_0) \sqsubseteq_S \mathrm{Cl}_S(X_1) \sqsubseteq_S \cdots$ of such sets, is a domain with basis $\{\, \mathrm{Cl}_S(X) \mid X \in M(D)\,\}$. It is isomorphic to the Smyth powerdomain of D. If D is an SFP domain, so is $\mathcal{P}_S(D)$.*

Let $\mathrm{Cl}_S(X_0) \sqsubseteq_S \mathrm{Cl}_S(X_1) \sqsubseteq_S \cdots$ be a chain of compact elements. As is shown in [Zh91], its least upper bound is the set $\mathrm{Cl}_S(SUP)$, where SUP is the collection of all least upper bounds of sequences $u_0 \sqsubseteq u_1 \sqsubseteq \cdots$ with $u_i \in X_i$.

Theorem 3.2. *Let D be a domain. Then the structure $(\mathcal{P}_H(D), \sqsubseteq_H)$, where $\mathcal{P}_H(D)$ consists of all sets $\mathrm{Cl}_H(X)$ with $X \in M(D)$ together with all unions $\bigcup_{i\in\omega} \mathrm{Cl}_H(X_i)$ of chains $\mathrm{Cl}_H(X_0) \sqsubseteq_H \mathrm{Cl}_H(X_1) \sqsubseteq_H \cdots$ of such sets, is a domain with basis $\{\, \mathrm{Cl}_H(X) \mid X \in M(D)\,\}$. It is isomorphic to the Hoare powerdomain of D. If D is an SFP domain, so is $\mathcal{P}_H(D)$.*

It is not possible to characterize the least upper bounds of increasing chains in $\mathcal{P}_H(D)^0$ as in the case of the Smyth powerdomain. In order to see this, consider the set $\omega \cup \{\omega\}$, which is a domain with respect to its natural order. Then the sequence $(\mathrm{Cl}_H(\{i\}))_{i\in\omega}$ is increasing with respect to $\sqsubseteq_H$. Its least upper bound is ω. But the collection of all least upper bounds of increasing sequences $(u_i)_{i\in\omega}$ with $u_i \in \mathrm{Cl}_H(\{i\})$ is $\omega \cup \{\omega\}$.

Note that the construction of a domain of subsets of D isomorphic to the Plotkin powerdomain in [Zh91] is based on the assumption that such a characterization can be given. However, following an idea of Plotkin [Pl83] and using the operator Cl_P as defined above a similar result can be obtained (cf. [Gr98]).

For an increasing chain $\mathrm{Cl}_P(X_0) \sqsubseteq_P \mathrm{Cl}_P(X_1) \sqsubseteq_P \cdots$ let $\mathrm{SUP}(X_i)_{i\in\omega}$ be the set of least upper bounds of all sequences $u_0 \sqsubseteq u_1 \sqsubseteq \cdots$ with $u_i \in X_i$.

Theorem 3.3. *Let D be a domain. Then the structure $(\mathcal{P}_P(D), \sqsubseteq_P)$, where $\mathcal{P}_P(D)$ consists of all sets $\mathrm{Cl}_P(X)$ with $X \in M(D)$ together with* *all sets* $\mathrm{Cl}_L(\mathrm{SUP}(X_i)_{i\in\omega})$*, where* $\mathrm{Cl}_P(X_0) \sqsubseteq_P \mathrm{Cl}_P(X_1) \sqsubseteq_P \cdots$ *is a chain of such elements, is a domain with basis* $\{\mathrm{Cl}_P(X) \mid X \in M(D)\}$*. It is isomorphic to the Plotkin powerdomain of* D*. If* D *is an SFP domain, so is* $\mathcal{P}_P(D)$*.*

4. Strongly finite sequent structures

Sequent structures have been introduced by Zhang [Zh91] as an analogue to Scott's information systems [Sc82] that captures exactly the SFP domains. A sequent structure consists of a set of propositions together with an entailment relation between finite sets of propositions satisfying the structural rules of Gentzen's sequent calculus for propositional logic.

In Scott's original definition every information system contained a distinguished proposition which could be thought of as being a tautology. In later papers on information systems such as [LW84] this condition has been omitted. Here, we modify Zhang's notion of a sequent structure by requiring the existence of such a distinguished element. The advantage of doing so will become clear in what follows.

Definition 4.1. A *sequent structure* $(A, T, \vdash)$ consists of a countable set A, the elements of which are called *propositions* or *tokens*, a distinguished element T of A, and a relation $\vdash$ between finite subsets of A, called *entailment* relation, which satisfies the following conditions:

(*Identity*) $\quad a \vdash a,$

(*Weakening*) $$\frac{X' \supseteq X \qquad X \vdash Y \qquad Y \subseteq Y'}{X' \vdash Y'},$$

(*Cut*) $$\frac{X \vdash Y, a \qquad a, X' \vdash Y'}{X, X' \vdash Y, Y'},$$

(*Truth*) $\quad \emptyset \vdash Y \Leftrightarrow T \in Y.$

Note that we write a singleton $\{a\}$ as a, and $X, X' \vdash Y, Y'$ for $X \cup X' \vdash Y \cup Y'$.

A sequent structure determines a collection of sets of propositions called its elements.

Definition 4.2. Let $\underline{A} = (A, T, \vdash)$ be a sequent structure. An *element* of $\underline{A}$ is a subset x of A closed under entailment:

$$X \subseteq x \wedge X \vdash Y \Rightarrow x \cap Y \neq \emptyset.$$

The set of all elements of $\underline{A}$ is denoted by $|\underline{A}|$.

Observe that by the truth requirement T is contained in any element. Moreover, the set $\{T\}$ is itself an element, since if X is a subset of $\{T\}$, then either X is empty or $X = \{T\}$. Let Y be a set of tokens. If $\emptyset \vdash Y$ then $T \in Y$, and if $T \vdash Y$ then also $\emptyset \vdash Y$, by the cut rule, as $\emptyset \vdash T$. Hence, $T \in Y$ also in this case.

Theorem 4.1. *Let $\underline{A} = (A, T, \vdash)$ be a sequent structure. Then $|\underline{A}|$ ordered by inclusion is a cpo with least element $\{T\}$.*

Note that this cpo need not be algebraic, which will, however, be the case under two further requirements. Let $(A, T, \vdash)$ be a sequent structure. For finite subsets X, Y of A define

$$X \vdash_\wedge Y \Leftrightarrow (\forall b \in Y) X \vdash b,$$
$$X \approx Y \Leftrightarrow X \vdash Y \wedge (\forall b \in Y) b \vdash_\wedge X.$$

Observe that $\approx$ is not symmetric. We say that a subset Y of A is $\approx$-*closed* if for every nonempty subset Y' of Y there is a subset Y'' of Y with $Y' \approx Y''$.

Definition 4.3. A sequent structure $(A, T, \vdash)$ is *strongly finite* if the following two conditions hold:

Finite closure: Every finite nonempty subset of A has a $\approx$-closed finite superset in A.

Atomicity: For every $a \in A$ and $X \subseteq A$ with $a \vdash X$ there is some $b \in X$ such that $a \vdash b$.

To see that these two requirements are not fulfilled in general consider the sequent structure $(\{a, b, T\}, T, \vdash)$ where $\vdash$ is generated by $a \vdash a$, $b \vdash b$ and $\emptyset \vdash T$. Then the axiom of finite closure does not hold since there is no subset Y of $\{a, b, T\}$ with $\{a, b\} \approx Y$. If, e.g., one had that also $\{a, b\} \vdash \emptyset$, then the resulting sequent structure would be strongly finite.

As in [Zh91] it can now be shown that for strongly finite sequent structures $\underline{A}$, $|\underline{A}|$ is an SFP domain with compact elements $\bar{a}$ $(a \in A)$, where $\bar{a} = \{\, b \in A \mid a \vdash b \,\}$. We will, however, present a different way to see this.

5. Sequent structures and preorders

Let $(A, T, \vdash)$ be a sequent structure and for tokens $a, b \in A$, let $a \precsim b$ if $a \vdash b$. Then $(A, \precsim)$ is a countable preorder with T as greatest element, which means that $a \precsim T$, for all $a \in A$, and if $T \precsim a$ then $a = T$. We say that $(A, \precsim)$ is the preorder *associated* with $(A, T, \vdash)$. As we will see next, any countable preorder with greatest element defines a sequent structure.

Proposition 5.1. *Let $(A, \precsim)$ be a countable preorder with greatest element 1 and for finite subsets X, Y of A define*

$$X \vdash Y \Leftrightarrow (\forall a \in A)[(\forall c \in X) a \precsim c \Rightarrow (\exists d \in Y) a \precsim d].$$

Then $(A, 1, \vdash)$ is a sequent structure for which the atomicity requirement holds.

We say that $(A, 1, \vdash)$ is the sequent structure *generated* by $(A, \precsim)$.

As we have just seen, $(A, 1, \vdash)$ is nearly strongly finite. Let us see now which condition we have to impose on $(A, \precsim)$ so that $(A, 1, \vdash)$ satisfies the finite closure requirement.

Definition 5.1. Let $(A, \precsim)$ be a countable preorder and X, Y be subsets of A. We say and write, respectively:

1 Y is *near* to X if above every lower bound of X there is an element of Y.

2 $X \rightleftharpoons Y$ if Y is a set of lower bounds of X that is near to X.

3 X is $\rightleftharpoons$-closed if for every nonempty subset Y' of X there is a subset Y'' of X with $Y' \rightleftharpoons Y''$.

Note that if A has a greatest element 1, then $\emptyset \rightleftharpoons Y$ exactly if $1 \in Y$. $(A, \precsim)$ is said to have the *nearness property* if every finite nonempty subset X of A has a finite $\rightleftharpoons$-closed superset in A.

Now, let $(A, 1, \vdash)$ be the sequent structure generated by $(A, \precsim)$. Then $X \vdash Y$ just if Y is near to X. Moreover, $b \vdash_\wedge X$ if and only if b is a lower bound of X.

Corollary 5.2. *Let $(A, \precsim)$ be a countable preorder with greatest element 1. Then the sequent structure generated by it is strongly finite exactly if $(A, \precsim)$ has the nearness property.*

Let PN be the collection of all countable preorders with greatest element that have the nearness property and SFSS be the collection of all strongly finite sequent structures. Moreover, let $\Psi\colon \mathrm{PN} \to \mathrm{SFSS}$ be the mapping which maps a preorder in PN to the sequent structure generated by it. As we will show next, this map is a bijection, its inverse being the mapping Ω that maps a strongly finite sequent structure to its associated preorder.

Theorem 5.3. *Let* $(A, T, \vdash)$ *be a strongly finite sequent structure and* $(A, \precsim)$ *be its associated preorder. Moreover, let* X, Y *be finite subsets of* A. *Then*

$$X \vdash Y \Leftrightarrow (\forall a \in A)[(\forall c \in X) a \precsim c \Rightarrow (\exists d \in Y) a \precsim d].$$

Proof. Suppose that $X \vdash Y$ and let $a \in A$ such that $a \precsim c$, for all $c \in X$. It follows that $a \vdash Y$ (cf. [Zh91, Theorem 3.1(2)]). Hence, there is some $d \in Y$ with $a \vdash d$, by the atomicity requirement.

For the proof of the converse implication consider first the case that X is empty. Then T is a lower bound of X. Thus, there is some $d \in Y$ with $T \precsim d$, which implies that $d = T$. This shows that $T \in Y$. By the truth requirement we therefore have that $\emptyset \vdash Y$.

If X is not empty, then, because of the finite closure condition, there is some subset Y' of A with $X \approx Y'$, which implies that $X \vdash Y'$ and $b \vdash c$, for all $b \in Y'$ and $c \in X$. By our assumption it follows for $b \in Y'$ that $b \vdash d$, for some $d \in Y$. Thus $b \vdash Y$, for every $b \in Y'$. With repeated applications of the cut rule we therefore obtain that $X \vdash Y$. □

It follows that the sequent structure generated by the preorder associated with $(A, T, \vdash)$ coincides with $(A, T, \vdash)$. With Corollary 5.2 we therefore obtain the following consequence.

Corollary 5.4. *Let* $(A, T, \vdash)$ *be a strongly finite sequent structure. Then its associated preorder has the nearness property.*

This shows that Ω is indeed a mapping from SFSS to PN. Since, by Proposition 5.1, the preorder associated with the sequent structure generated by a given preorder $(A, \precsim)$ with greatest element coincides with $(A, \precsim)$, we have that Ω and Ψ are invers to each other. As we will see next, SFSS and PN are large cpo's with respect to the substructure ordering and both maps are continuous.

Definition 5.2. Let $\underline{A} = (A, T_A, \vdash_A)$ and $\underline{B} = (B, T_B, \vdash_B)$ be strongly finite sequent structures. $\underline{A} \trianglelefteq \underline{B}$, read $\underline{A}$ is a *substructure* of $\underline{B}$, if $A \subseteq B$ and for all finite subsets X, Y of A,

$$X \vdash_A Y \Leftrightarrow X \vdash_B Y.$$

Note that if $\underline{A} \trianglelefteq \underline{B}$ then $T_A = T_B$, by the truth condition. Without restriction we identify all distinguished tokens and assume that they are all denoted by T.

Proposition 5.5. $(\mathrm{SFSS}, \trianglelefteq)$ *is a large algebraic cpo with the finite strongly finite sequent structures as compact elements.*

Proof. As in [Zh91] we obtain that $(\mathrm{SFSS}, \trianglelefteq)$ is a large cpo. For the proof of algebraicity let $\underline{A}$ be in SFSS and X be a finite nonempty subset of A that contains T. By strong finiteness of $\underline{A}$ there is a finite $\approx_A$-closed superset Y of X. Define $\vdash_Y$ by restricting $\vdash_A$ to the powerset of Y. Then $\underline{Y} = (Y, T, \vdash_Y)$ is in SFSS with $\underline{Y} \trianglelefteq \underline{A}$. Since the least upper bound of a directed collection is given by the union of the structures in the collection, it is now readily verified that $\{\, \underline{Y} \in \mathrm{SFSS} \mid Y \text{ finite} \wedge \underline{Y} \trianglelefteq \underline{A} \}$ is directed with least upper bound $\underline{A}$. □

Definition 5.3. Let $\widetilde{A} = (A, \precsim_A, 1_A)$ and $\widetilde{B} = (B, \precsim_B, 1_B)$ be preorders in PN. $\widetilde{A} \unlhd \widetilde{B}$, read $\widetilde{A}$ is a *subpreorder* of $\widetilde{B}$, if $A \subseteq B$, $1_A = 1_B$, $\precsim_A = \precsim_B \cap (A \times A)$, and for all finite nonempty subsets X of A and all $m \in B$,

$$(\forall a \in X) m \precsim_B a \Rightarrow (\exists n \in A) m \precsim_B n \wedge (\forall a \in X) n \precsim_A a.$$

In this case we identify all greatest elements and denote them by 1.

Proposition 5.6. $(\mathrm{PN}, \unlhd)$ *is a large algebraic cpo with the finite preorders in PN as compact elements.*

The proof of this result parallels that for Proposition 5.5.

Lemma 5.7. *The maps* $\Psi \colon \mathrm{PN} \to \mathrm{SFSS}$ *and* $\Omega \colon \mathrm{SFSS} \to \mathrm{PN}$ *are monotone.*

Proof. We only show that Ω is monotone. Let to this end $\underline{A}$, $\underline{B}$, respectively, be strongly finite sequence structures with $\underline{A} \trianglelefteq \underline{B}$ and $\widetilde{A}, \widetilde{B}$ be their associated preorders. We have to show that $\widetilde{\underline{A}} \unlhd \widetilde{\underline{B}}$. We will only verify the last condition in Definition 5.3. Let to this end X be a finite nonempty subset of A and $m \in B$ such that $m \precsim_B a$, for all $a \in X$. Since $\underline{A}$ satisfies the finite closure condition, there is some finite subset Y of A such that $X \approx_A Y$. Thus $X \vdash_A Y$ and $b \precsim_A a$, for all $b \in Y$ and $a \in X$. It follows that also $X \vdash_B Y$, from which we obtain by Theorem 5.3 that there is some $n \in Y$ with $m \precsim_B n$. Because Y is contained in A, we obtain that $n \in A$ and $n \precsim_A a$, for all $a \in X$. □

As is readily seen, Ψ and Ω are even continuous.

Theorem 5.8. *The large cpo's* $(\mathrm{SFSS}, \trianglelefteq)$ *and* $(\mathrm{PN}, \unlhd)$ *are isomorphic.*

Because of the neat relationship between strongly finite sequent structures and preorders with greatest element that have the nearness property constructions on one of theses classes can easily be translated in constructions on the other. In what follows we will work on the preorder side. Let us first define when two preorders are isomorphic.

Definition 5.4. Let $\widetilde{A}, \widetilde{B}$ be preorders. A map $f\colon A \to B$ is a *preisomorphism* if it is surjective, order-preserving and order-reflecting.

Next, we will study what kind of subsets of a preorder $\widetilde{A} \in \mathrm{PN}$ correspond to the elements determined by the sequence structure generated by $\widetilde{A}$. Recall that a subset of A is a filter if it is upwards closed with respect to the preorder and contains a lower bound of any of its finite subsets.

Proposition 5.9. *Let $\widetilde{A} = (A, \precsim)$ be a preorder in PN and $\underline{A} = (A, 1, \vdash)$ be the strongly finite sequent structure generated by it. Then $x \in |\underline{A}|$ if and only if x is a filter in $\widetilde{A}$.*

Proof. Let $x \in |\underline{A}|$. Then x is upwards closed with respect to $\precsim$. Moreover, if X is a finite subset of A there is a finite subset Y of A so that $X \vdash Y$ and $b \precsim c$, for all $b \in Y$ and $c \in X$. Since $x \in |\underline{A}|$, there is some $a \in Y \cap x$. Hence, a is a lower bound of X.

Conversely, if x is a filter in $\widetilde{A}$ and X is a finite subset of x with $X \vdash Y$, then x contains a lower bound of X. Since $\underline{A}$ is generated by $\widetilde{A}$, we have that there is some $b \in Y$ with $a \precsim b$. It follows that $b \in x$. □

Let $\precsim^*$ be the dual preorder to $\precsim$. Then we obtain that $x \in |\underline{A}|$ exactly if x is an ideal with respect to $\precsim^*$. Thus, the construction of an SFP domain from a given strongly finite sequent structure corresponds to the ideal completion of the preorder which is dual to the preorder associated with the given sequent structure.

For a countable preorder $\widetilde{A}$ let $\mathcal{F}(\widetilde{A})$ be the set of its filters. Then $(\mathcal{F}(\widetilde{A}), \subseteq)$ is a domain with the principal filters as its compact elements. The next lemma shows how the relation $\rightleftharpoons$ is related to the computation of minimal upper bounds in $\mathcal{F}(\widetilde{A})$.

Lemma 5.10. *Let $\widetilde{A}$ be a countable preorder with greatest element and X, Y be finite subsets of A with $X \rightleftharpoons Y$. Then $\{\uparrow\{b\} \mid b \in Y \wedge b \text{ maximal}\} = \mathrm{MUB}(\{\uparrow\{a\} \mid a \in X\})$.*

Proof. Clearly, each filter $\uparrow\{b\}$ with $b \in Y$ is an upper bound of $\{\uparrow\{a\} \mid a \in X\}$. Let $b \in Y$ and $y \in \mathcal{F}(\widetilde{A})$ with $y \subseteq \uparrow\{b\}$ be a further upper bound of $\{\uparrow\{a\} \mid a \in X\}$. Then $X \subseteq y$. It follows that X has a lower bound c in y. Since $X \rightleftharpoons Y$, there is some $d \in Y$ with $c \precsim d$. Thus $d \in y$, which implies that $b \precsim d$. As b is maximal, we obtain that also $d \precsim b$. This shows that $\uparrow\{b\} \subseteq y$ as well.

Now, conversely, let $\uparrow\{c\} \in \text{MUB}(\{\uparrow\{a\} \mid a \in X\})$. Then c is a lower bound of X. Hence, there is some $d \in Y$ with $c \precsim d$. Because $d \in Y$, $\uparrow\{d\}$ is an upper bound of $\{\uparrow\{a\} \mid a \in X\}$. Moreover, $\uparrow\{d\} \subseteq \uparrow\{c\}$. By the minimality of $\uparrow\{c\}$ it follows that $\uparrow\{d\} = \uparrow\{c\}$ as well as that d is maximal in Y. □

For an SFP domain $(D, \sqsubseteq)$ let $\mathcal{S}(D)$ be the partial order $(D^0, \sqsubseteq^*)$, where $\sqsubseteq^*$ is the dual order to $\sqsubseteq$. Apparently, for any finite set Z of compact elements of D, $Z \rightleftharpoons_{\mathcal{S}(D)} \text{MUB}(Z)$.

Theorem 5.11. *1 Let $\widetilde{A}$ be a preorder in PN. Then $\mathcal{F}(\widetilde{A})$ is an SFP domain. Moreover, the mapping which maps each element of A to the principal filter generated by it is a preisomorphism from $\widetilde{A}$ onto $\mathcal{S}(\mathcal{F}(\widetilde{A}))$.*

2 Let D be an SFP domain. Then $\mathcal{S}(D)$ is in PN and the mapping that maps every filter in $\mathcal{F}(\mathcal{S}(D))$ to its greatest lower bound is an isomorphism from $\mathcal{F}(\mathcal{S}(D))$ onto D.

6. Constructions on preorders

We will now deal with the constructions on preorders which correspond to those on domains considered earlier. In the function space case we will present the construction of what corresponds to the full continuous function space of domains.

The definitions of the lift $\widetilde{A}_*$, the sum $\widetilde{A} \oplus \widetilde{B}$ and the product $\widetilde{A} \otimes \widetilde{B}$ of preorders $\widetilde{A}$, $\widetilde{B}$ in PN are completely analogous to those in the domain case and will be skipped. An straightforward verification shows that these preorders are in PN too. Moreover, $\mathcal{F}(\widetilde{A}_*)$, $\mathcal{F}(\widetilde{A} \oplus \widetilde{B})$ and $\mathcal{F}(\widetilde{A} \otimes \widetilde{B})$, respectively, are isomorphic to $\mathcal{F}(\widetilde{A})_\perp$, $\mathcal{F}(\widetilde{A}) + \mathcal{F}(\widetilde{B})$ and $\mathcal{F}(\widetilde{A}) \times \mathcal{F}(\widetilde{B})$.

The *Smyth, Hoare* and *Plotkin power preorders* $P_S(\widetilde{A})$, $P_H(\widetilde{A})$ and $P_P(\widetilde{A})$, respectively, consist of the collection $M(A)$ of finite nonempty subsets α of A which either only contain the element 1_A or do not contain it and which are preordered by

$$\alpha \precsim_S \beta \Leftrightarrow (\forall a \in \alpha)(\exists b \in \beta) a \precsim_A b,$$
$$\alpha \precsim_H \beta \Leftrightarrow (\forall b \in \beta)(\exists a \in \alpha) a \precsim_B b,$$
$$\alpha \precsim_P \beta \Leftrightarrow \alpha \precsim_S \beta \wedge \alpha \precsim_H \beta.$$

Theorem 6.1. *Let $\widetilde{A}$ be a preorder in PN. Then $P_S(\widetilde{A})$, $P_H(\widetilde{A})$ and $P_P(\widetilde{A})$ are in PN as well. Besides, $\mathcal{F}(P_S(\widetilde{A}))$, $\mathcal{F}(P_H(\widetilde{A}))$ and $\mathcal{F}(P_P(\widetilde{A}))$, respectively, are isomorphic to $\mathcal{P}_S(\mathcal{F}(\widetilde{A}))$, $\mathcal{P}_H(\mathcal{F}(\widetilde{A}))$ and $\mathcal{P}_P(\mathcal{F}(\widetilde{A}))$.*

Proof. Obviously, $P_S(\widetilde{A})$, $P_H(\widetilde{A})$ and $P_P(\widetilde{A})$ are preorders with greatest element $\{1_a\}$. We will now verify that they have the nearness property. In the case of the Smyth power preorder this follows as in [Zh91, Proposition 3.10].

Let us next consider the Hoare power preorder and let X be a finite nonempty subset of $M(A)$. Set $\beta_X = \bigcup X$. Then it is easily verified that $X \rightleftharpoons_H \{\beta_X\}$. Therefore, if V is a finite nonempty subset of $M(A)$ and $X_1, \ldots, X_k$ are its nonempty subsets, set $V_e = V \cup \{\beta_{X_1}, \ldots, \beta_{X_k}\}$. Then V_e is a $\rightleftharpoons_H$-closed superset of V. To see this, let Y' be a nonempty subset of V_e. Then there are nonempty subsets $X', X'_1, \ldots, X'_r$ of V so that $Y' = X' \cup \{\beta_{X'_1}, \ldots, \beta_{X'_r}\}$. Set $\beta = \bigcup X' \cup X'_1 \cup \cdots \cup X'_r$. Then $\beta \in V_e$ and $Y' \rightleftharpoons_H \{\beta\}$.

Finally, we consider the Plotkin power preorder. Let V be a finite nonempty subset of $M(A)$. Then there is a finite $\rightleftharpoons_A$-closed subset α of A which contains $\bigcup V$. Let Z be the set of all nonempty subsets of α. We will show that any nonempty subset X of Z has a $\rightleftharpoons_{P_P(\widetilde{A})}$-closed superset X' in Z. If X has no lower bounds, we let X' be the empty set. Otherwise, assume that $X = \{\, \alpha_i \mid i \in I \,\}$, where I is finite. For every choice function $f\colon I \to \bigcup X$ with $f(i) \in \alpha_i$, for $i \in I$, there is some subset β_f of α such that $\{\, f(i) \mid i \in I \,\} \rightleftharpoons_A \beta_f$. Let β be the union of all such sets β_f and define X' to be the collection of all subsets of β which intersect every β_f. As is easily verified, X' has the desired properties.

The remaining statement follows by the correspondence of the domain construction from preorders in PN used here and ideal completion. □

The appropriate concept of morphism on preorders in PN which corresponds to continuous maps on the associated domains is that of an approximable mapping.

Definition 6.1. Let $\widetilde{A}, \widetilde{B}$ be preorders in PN. An *approximable mapping* from $\widetilde{A}$ to $\widetilde{B}$ is a relation $R \subseteq A \times B$ which satisfies the following two conditions:

1 $aRb \wedge b \precsim_B b' \Rightarrow aRb'$.

2 For all finite subsets S of R and all subsets X, Y of A and B, respectively,

$$p_1(S) \rightleftharpoons_A X \wedge p_2(S) \rightleftharpoons_B Y \Rightarrow (\forall a \in X)(\exists b \in Y) aRb.$$

Since $\emptyset \rightleftharpoons_A \{1_A\}$, the case that S is empty implies that $1_A R 1_B$. The next result follows essentially as in [Zh91].

Proposition 6.2. *Let $\widetilde{A}, \widetilde{B}$ be preorders in PN and R be an approximable mapping from $\widetilde{A}$ to $\widetilde{B}$. Define $f_R\colon \mathcal{F}(A) \to \mathcal{F}(B)$ by*

$$f_R(x) = \{\, b \in B \mid (\exists a \in x) aRb \,\}.$$

Then f_R is continuous.

As has been mentioned earlier, the compact elements of the space of all continuous functions between two SFP domains are those functions that are generated by joinable sets. Joinable sets are finite sets of pairs of compact elements, which in turn are principal filters generated by tokens of the preorders corresponding to the domains under consideration. If we remove the intermediate step of generating filters, we obtain the notion of a molecule.

Definition 6.2. Let $\widetilde{A}, \widetilde{B}$ be preorders on PN. A finite subset m of $A \times B$ is a *molecule* if the following three conditions hold:

1 $(1_A, 1_B) \in m$.

2 $p_2(m) = \{1_B\} \Rightarrow p_1(m) = \{1_A\}$.

3 For every subset m' of m there are sets $X \subseteq p_1(m)$ and Y such that their elements are incomparable with respect to the preorder and

$$p_1(m') \rightleftharpoons_A X \wedge p_2(m') \rightleftharpoons_B Y \wedge (\forall a \in X)(\exists b \in Y)(a, b) \in m.$$

Let $\widetilde{A}, \widetilde{B}$ be preorders in PN. Then the *function space* $\widetilde{A} \to \widetilde{B}$ is the set of all molecules in $A \times B$ preordered by

$$m \precsim m' \Leftrightarrow (\forall (a, b) \in m')(\exists (c, d) \in m) a \precsim_A c \wedge d \precsim_B b.$$

The greatest element is $\{(1_A, 1_B)\}$.

To see that $\{(1_A, 1_B)\}$ is indeed a molecule choose $X = \{1_A\}$ and $Y = \{1_B\}$ in Definition 6.2.

Proposition 6.3. *Let $\widetilde{A}, \widetilde{B}$ be preorders in PN and m be a molecule in $\widetilde{A} \to \widetilde{B}$. Define $R_m \subseteq A \times B$ by*

$$aR_m b \Leftrightarrow (\exists (c, d) \in m) a \precsim_A c \wedge d \precsim_B b.$$

Then R_m is an approximable mapping from $\widetilde{A}$ to $\widetilde{B}$. Moreover, for $m', m'' \in \widetilde{A} \to \widetilde{B}$,

$$m' \precsim m'' \Leftrightarrow f_{R_{m''}} \sqsubseteq f_{R_{m'}}.$$

Proof. Let S be a finite subset of R_m and X, Y be such that $p_1(S) \rightleftharpoons_A X$ and $p_2(S) \rightleftharpoons_B Y$. We consider only the case that S is not empty. Set

$$m' = \{ (a, b) \in m \mid (\exists (s, t) \in S) s \precsim_A a \wedge b \precsim_B t \}.$$

It follows that there are sets X', Y' with $X' \subseteq p_1(m)$, $p_1(m') \rightleftharpoons_A X'$ and $p_2(m') \rightleftharpoons_B Y'$ so that for any $a \in X'$ there is some $b \in Y'$ with $(a, b) \in m$. Let $c \in X$. Then c is a lower bound of $p_1(S)$ and hence a lower bound of $p_1(m')$, which implies that $c \precsim_A a$, for an $a \in X'$. Thus, there is some $b \in Y'$ with $(a, b) \in m$. Since $p_2(m') \rightleftharpoons_B Y'$, we obtain that b is a lower bound $p_2(m')$. It follows that $b \precsim_B d$, for some $d \in Y'$. This shows that there is a pair $(a, b) \in m$ with $c \precsim_A a$ and $b \precsim_B d$, i.e. $(c, d) \in R_m$. □

Lemma 6.4. *Let $\widetilde{A}, \widetilde{B}$ be preorders in PN and m be a molecule in $\widetilde{A} \to \widetilde{B}$. Set*

$$T_m = \{ (\uparrow\{a\}, \uparrow\{b\}) \mid (a, b) \in m \}.$$

Then T_m is joinable with generated step function f_{R_m}.

The joinability is a consequence of Lemma 5.10 and the last requirement in the definition of a molecule. A routine calculation shows moreover that f_{R_m} is the step function generated by T_m.

Lemma 6.5. *Let $\widetilde{A}, \widetilde{B}$ be preorders in PN and $J \subseteq \mathcal{F}(\widetilde{A})^0 \times \mathcal{F}(\widetilde{B})^0$ be a joinable set such that $p_2(J) \neq \{\uparrow\{1_A\}\}$. Set*

$$m_J = \{ (a, b) \in A \times B \mid (\uparrow\{a\}, \uparrow\{b\}) \in J \}.$$

Then m_j is a molecule in $\widetilde{A} \to \widetilde{B}$ with $f_{R_{m_J}}$ being the step function generated by J.

The verification that m_J satisfies the conditions in Definition 6.2 is routine. Since $T_{m_J} = J$ it follows from the preceding lemma that $f_{R_{m_J}}$ is the step function generated by J.

As a consequence of these results we have for preorders $\widetilde{A}, \widetilde{B}$ in PN that the mapping which maps molecules m in $\widetilde{A} \to \widetilde{B}$ to the function f_{R_m} induces an order-isomorphism between the partially ordered sets $\{ \uparrow\{m\} \mid m \in \widetilde{A} \to \widetilde{B} \}$ and $[\mathcal{F}(\widetilde{A}) \to \mathcal{F}(\widetilde{B})]^0$, which extends to an isomorphism of the domains $\mathcal{F}(\widetilde{A} \to \widetilde{B})$ and $[\mathcal{F}(\widetilde{A}) \to \mathcal{F}(\widetilde{B})]$. It follows that $\mathcal{F}(\widetilde{A} \to \widetilde{B})$ is an SFP domain and hence $\widetilde{A} \to \widetilde{B}$ is in PN.

Theorem 6.6. *Let $\widetilde{A}, \widetilde{B}$ be preorders in PN. Then $\widetilde{A} \to \widetilde{B}$ is in PN as well. Moreover, $\mathcal{F}(\widetilde{A} \to \widetilde{B})$ is isomorphic to $[\mathcal{F}(\widetilde{A}) \to \mathcal{F}(\widetilde{B})]$.*

7. Conclusion

Strongly finite sequent structures have been introduced in [Zh91] as a logic-oriented representation of SFP domains. A sequent structure is a set of propositions with an entailment relation between finite sets of propositions. In this paper, following a suggestion of the referee, sequence structures having a distinguished proposition with the properties of a tautology are considered. Doing so is not restriction compared with the approach in [Zh91], but has certain advantages. E.g. the representation of the compact elements of the domain generated by a sequence structure becomes more uniform.

Moreover, it is shown in this paper that there is a close connection between sequent structures and preorders with greatest element. In the strongly finite case every entailment relation can be derived from a preorder. As follows from the results in this paper, preorders are an intermediate concept between sequence structures and domains, which helps to clarify their relationship. To each of the domain constructions used in programming language semantics there is a similar construction for preorders from which the corresponding construction for sequent structures can be obtained in a uniform and simple way.

Acknowledgement

The authors are grateful to the unknown referee for his valuable remarks.

References

[Ab91] Abramksy, S., Domain theory in logical form, *Annals of Pure and Applied Logic* **51** (1991) 1–77.

[AJ94] Abramsky, S. and Jung, A., Domain Theory, In: S. Abramsky et al. (editors), *Handbook of Logic in Computer Science, Vol. 3, Semantic Structures*, Oxford University Press, 1994, 1–168.

[AC98] Amadio, R. M. and Curien, P.-L., *Domains and Lambda Calculi*, Cambridge University Press, 1998.

[Gr98] Greb, R., The category of strongly finite sequent structures, Informatik-Bericht Nr. 02-98, University of Siegen, 1998.

[Gu92] Gunter, C., *Semantics of Programming Languages*, MIT Press, 1992.

[GS90] Gunter,C. and Scott, D. S., Semantic domains, In: J. van Leeuven (editor), *Handbook of Theoretical Computer Science, Vol. B, Formal Methods and Semantics*, Elsevier, 1990, 633–674.

[Ju89] Jung, A., Cartesian closed categories of domains, CWI Tracts **66**, Centrum voor Wiskunde en Informatica, Amsterdam, 1989.

[LW84] Larsen, K. G. and Winskel, G., Using information systems to solve recursive domain equations effectively, In: G. Kahn et al. (editors), *Semantics of Data Types*, Lecture Notes in Computer Science **173**, Springer, 1984, 109–130.

[Pl76] Plotkin, G. D., A powerdomain construction, *SIAM J. Computing* **5** (1976) 452–487.

[Pl83] Plotkin, G. D., Domains, Department of Computer Science, University of Edinburgh, 1983.

[Sc70] Scott, D. S., Outline of a mathematical theory of computation, Technical Monograph PR-2, Oxford University Computation Laboratory, 1970.

[Sc82] Scott, D. S., Domains for denotational semantics, In: M. Nielson et al. (editors), *International Colloquium on Automata, Languages and Programming*, Lecture Notes in Computer Science **140**, Springer, 1982, 577–613.

[SLG94] Stoltenberg-Hansen, V., Lindström, I. and Griffor, E. R., *Mathematical Theory of Domains*, Cambridge University Press, 1994.

[Zh91] Zhang, G. Q., *Logic of Domains*, Birkhäuser, 1991.

SEMANTIC STRUCTURES IN COMPUTATION

1. K. Keimel, GQ Zhang, Y.-M. Liu and Y. Chen (eds.): *Domains and Processes.* Proceedings of the 1st International Symposium on Domain Theory. 2001
ISBN 0-7923-7143-7
2. P. Blain Levy: *Call-By-Push-Value.* A Functional/Imperative Synthesis. 2003
ISBN 1-4020-1730-8
3. GQ Zhang, J. Lawson, Y.-M. Liu and M.-K. Luo (eds.): *Domain Theory, Logic and Computation.* Proceedings of the 2nd International Symposium on Domain Theory, Sichuan, China, October 2001. 2003 ISBN 1-4020-1832-0

KLUWER ACADEMIC PUBLISHERS – DORDRECHT / BOSTON / LONDON

MIX
Papier aus verantwortungsvollen Quellen
Paper from responsible sources
FSC® C105338

If you have any concerns about our products,
you can contact us on
ProductSafety@springernature.com

In case Publisher is established outside the EU,
the EU authorized representative is:
Springer Nature Customer Service Center GmbH
Europaplatz 3, 69115 Heidelberg, Germany

Printed by Libri Plureos GmbH
in Hamburg, Germany